The International Space Station
From Imagination to Reality

FOREWORD

The International Space Station is already one of the brightest objects in the night sky with only half of the construction phase completed. It really is a project which both inspires and bores people.

The inspiration is the fact that so many nations have come together to use their expertise to construct this station. America and Russia have surrendered their individual aspirations of manned space flight and buried for ever the rivalry that marked the manned programmes of both countries from the dawn of human space flight. Astronauts and cosmonauts train together for years to work together in orbit. They are joined on crews by colleagues from the Space Agencies of Europe, Canada and Japan. All these agencies also offer their training facilities and provide hardware for the station. The crews work in orbit for months at a time. There are problems: over access, money, priority but this is what working together means. This is IMAGINATION in practice.

The boredom is that one day on the station seems like any other. The big headlines have gone. The every day grind of repair, construction, experiment and resupply has taken over. This is the REALITY of Spaceflight. It is what many people within the various space agencies have worked for as they want to make space travel an everyday event.

This is the first book of what we hope will be a series telling the story of how the International Space Station was established in orbit. This book looks at the early concepts which had to be discarded as money and complexities of construction made planning difficult. We have broken down the first series of missions into various compartments. The BIS asked some of the best writers on space to contribute to this volume. They are drawn from Europe, America and Great Britain.

We hope you will have in one place an essential guide to the key events of the first phase of construction up to December 2001. The British Interplanetary Society's motto is 'From Imagination to Reality'. It is appropriate in our 70th anniversary year that the ISS is also making that dream very real in the night sky.

I wish to thank all the contributors and the staff of the BIS who have made the task of editing this volume so enjoyable.

Rex Hall

The International Space Station
From Imagination to Reality

EDITOR: REX HALL

CONTENTS

From Mir-2 to the ISS Russian Segment   3
Bart Hendrickx

The International Space Station resulted from the amalgamation of the US/European/Canadian/Japanese Freedom station and Russia's Mir-2 station. This article will explore the evolution of Mir-2 and how it was incorporated into the ISS. It will also take a look at the development of the various Russian ISS modules and the numerous transformations that many of them have undergone in the past decade.

ISS American Elements On-Orbit Through Decemebr 2001   45
Roelof Schuiling

This chapter discusses the American ISS components launched and assembled up to December 2001. These include the Unity Node; Pressurized Mating Adapters -1, -2 and -3; the Destiny laboratory; Quest airlock and the Z-1 truss.

The International Space Station: Expedition Crews   56
Neville Kidger

Astronaut and cosmonaut crews began the permanent staffing of the International Space Station early in November 2000 after years of delay. Neville Kidger describes the activities of the first expedition crew using first-hand accounts from the Commander, William Shepherd, to illustrate the sometimes fraught difficulties encountered by the first trio of pioneers on the ISS. The major activities of the second and third crews are covered with emphasis upon the engineering and historic aspects of these expeditions.

NASA Shuttle Missions to ISS   81
David J. Shayler

This article reviews the operational accomplishments of the twelve American Space Shuttle assembly and re-supply missions flown between December 1998 and November 2001, which enabled the International Space Station to finally become a reality.

Research in Orbit   119
Andrew Salmon

Tommy Holloway, ISS programme manager, said at a press conference at the start of 2001: "…a rigorous science programme (on ISS) will not begin until the station is ready for a seven-person crew around 2005". Despite those words, what follows is an introduction to the wide variety of research carried out aboard the nascent ISS. The account ends with the handover of Expedition 3 to Expedition 4 in December 2001.

ISS Soyuz Crewing 2000-2001   144
Bert Vis

The first expedition crew was launched to ISS on a Russian Soyuz, a ship that from then on was to act as "life-boat". Given the limited lifetime of around 200 days, two replacement flights per year were needed. The crewing history of these flights is covered in this chapter, including the new phenomenon: space tourism.

The International Space Station - Orbital Considerations and Related Topics   149
Phillip Clark

This Chapter is built around two main tables. Table 1 provides a listing of all of the orbital manoeuvres by the International Space Station, from the launch of the first element, Zarya, in November 1998 through to the end of 2001: additionally, the orbit if the ISS Complex at the time of the arrival of each spacecraft docking is listed. Table 2 provides a listing of the main mission event dates/times for all of the spacecraft launched to ISS through to the end of 2001. As well as discussing issues connected specifically with these two Tables, the Chapter reviews anomalies in United States Space Command's policy of cataloguing ISS elements. The different systems of designating ISS missions are reviewed and finally orbital debris which has been tracked from ISS missions is discussed.

Conclusion   160
Rex Hall

From Mir-2 to the ISS Russian Segment

BART HENDRICKX

Introduction

Most of the originally planned elements of the ISS Russian segment were directly inherited from the final version of the Mir-2 space station. The major exception was the Russian-built and US-financed FGB/Zarya module, which despite a long heritage in the Almaz, Salyut and Mir programmes, was introduced specifically for ISS, mainly for political reasons. The FGB kicked off station assembly in November 1998 and after many frustrating delays was joined in July 2000 by the Zvezda Service Module, the original core of Mir-2. The only other element of the Russian segment launched thus far is the Pirs Docking Compartment, which linked up with the station in September 2001. Russia's seemingly never-ending financial problems forced numerous design changes in the remaining add-on elements, most of which will now be built through commercial partnerships with US companies. If the Russian segment is ever completed, it will look a whole lot different from what designers had envisaged in the early 1990s.

The Three Lives of Mir-2 [1]

Mir Back-up

The Base Block of the Mir space station, launched in February 1986, was an outgrowth of the civilian Salyut space stations. Officially known as the Longterm Orbital Stations (Russian acronym DOS), six of these had been launched between 1971 and 1982, four of which actually hosted crews. The first four stations were equipped with one docking port and the last two (Salyut-6 and 7) had two docking ports. They were all designed jointly by the Korolyov design bureau (known successively as OKB-1, TsKBEM, NPO Energiya and since 1994 as RKK Energiya) and a branch of the Chelomey bureau located in the Moscow suburb of Fili. Manufacturing took place at the Khrunichev factory, also situated in Fili.

In February 1976 the Soviet government gave the official go-ahead to begin the development of a next-generation space station with multiple docking ports. This called for the launch of two core modules, namely DOS-7K N°7 (serial number 17KS N° 12701) and DOS-7K N°8 (serial number 17KS N° 12801). Both were supposed to have a guaranteed lifetime of at least three years. This plan also provided some redundancy. In case DOS-7 failed to reach orbit or suffered some catastrophic in-orbit failure, DOS-8 could have been launched as a back-up. The same practice had been followed with all previous civilian Salyut stations, which were all manufactured in pairs to ensure the availability of a spare vehicle should the first one fail (DOS 1&2, DOS 3&4, DOS 5&5-2)

The original 1976 proposals for DOS-7 and DOS-8 involved a core module with a total of four docking ports, two axial ports on either end of the station and two lateral ports on the small-diameter work compartment. In 1978 it was decided to move the lateral docking ports to the front cylindrical transfer compartment and increase their number from two to four. After the Fili branch of Chelomey's bureau joined the project in 1979, the cylindrical transfer compartment was changed into a spherical docking adapter inherited from a cancelled successor of the military Almaz space stations known as Zvezda. In 1981 the Fili branch was incorporated into NPO Energiya and became known as the Salyut Design Bureau (KB Salyut). It separated from NPO Energiya in 1988 and eventually joined with the Khrunichev factory in 1993 to form the Khrunichev State Space Scientific Production Centre (GKNPTs Khrunichev) [2].

The final blueprints for DOS-7 and DOS-8 were finished in early 1982, making it possible for their construction to begin at the Khrunichev plant in Moscow. The hull of DOS-8 was finished in early February 1985. No other major work was done on it until the launch of DOS-7 in February 1986. If something had gone wrong with that launch, DOS-8 could have been sent up as a replacement within about 1 to 1.5 years.

The Giant Mir-2

Until the early 1980s plans for DOS-8 were either to replace DOS-7 in case of a launch failure or to succeed DOS-7 in orbit after it had outlived its usefulness, which was expected to be only a few years after launch. DOS-8 would have become the core of a space station very similar to the original Mir, with the most visible change being the addition of a large cross beam which would have been used to support solar panels and other equipment.

Plans for the Mir follow-on station changed in 1984, when NPO Energiya began devising plans for a giant orbital complex known as the Orbital Assembly and Operations Centre (Russian abbreviation OSETs for *Orbitalnyy sborochno-ekspluatatsionnyy tsentr*). Probably seen as a response to America's Freedom space station, it was supposed to be used primarily as an assembly site for large-scale orbital structures and a repair shop and refuelling station for orbiting satellites. DOS-8 was supposed to become the first element of OSETs and would be followed by several modules launched by the Energiya rocket. During the next two years the goals and timelines for the new station were revised. The build-up would be more gradual, with the actual goals of the OSETs being fulfilled only in the final stage. In December 1986 NPO Energiya came up with so-called Technical Proposals for such a station and on 14 December 1987 NPO Energiya's first deputy general designer Yuriy Semyonov approved the Draft Plan for the orbital complex, which in January 1988 was announced in the Soviet media as Mir-2. The designers also referred to it as 180GK.

According to the 1987 Draft Plan assembly of Mir-2/180GK was to get underway in August 1993 with the launch of DOS-8, which would act as the living quarters for the crew. The next launch, in October 1993, would have added a 76-tonne core module (MoB) to DOS-8 (Fig. 1). This was to be orbited by the 14A10 booster, a standard Energiya rocket with a new NPO Energiya upper stage based on the venerable Blok-DM, which was to provide the final push to orbit. An alternative option proposed by KB Salyut was the use of a module based on the Skif-DM/Polyus (17F19DM) payload flown on the first Energiya rocket and incorporating an FGB section.

The first phase of assembly was to be rounded out by June 1994 with the addition of a large truss to carry solar arrays (Fig. 2). The second phase of assembly (1994-1997) would see the launch of three more giant modules (a technological module, a support module and a biotechnological module) and the further expansion of the truss with solar arrays, solar parabolic concentrators and scientific platforms (Fig. 3). In the final phase (1997-2000) Mir-2/GK-180 would be expanded to fulfil the role of the OSETs. This involved the launch of four more giant modules and also the deployment of various types of space tugs. The station would also be equipped with a fuel storage depot and various facilities to repair satellites and assemble large structures in space.

The intended inclination of Mir-2/180GK was 65° to expand remote sensing coverage of Soviet territory. This inclination had also been foreseen for Mir, but one year before the launch of the DOS-7 core module it was changed

Fig. 1 One version of the Energiya-launched core module (MoB) of Mir-2/GK-180.
(source: RKK Energiya)

Key:
1. Platform for truss segments;
2. Platform attachment structure;
3. EVA exit hatch;
4. Folded solar panel;
5. Unfolded solar panel;
6. Visual observation post;
7. Attachment structure for truss;
8. Storage room for truss segments;
9. Remote manipulator arm;
10. Axial APAS docking port;
11. Rendezvous antennas;
12. Radial APAS docking port;
13. Propellant tanks;
14. Living compartment;
15. Technological compartment;
16. Transfer compartment;
17. Gyrodin section;
18. Approach and orientation engines;
19. Storage room for truss segments.

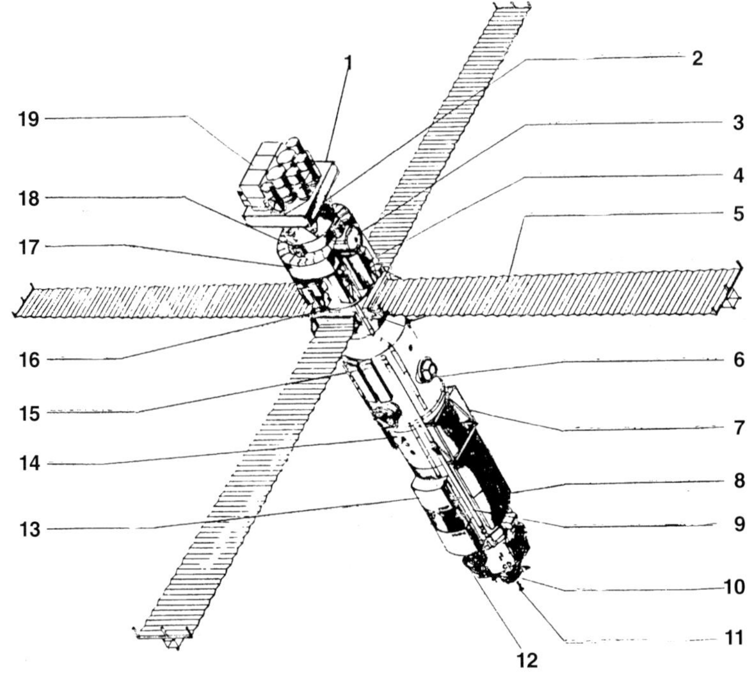

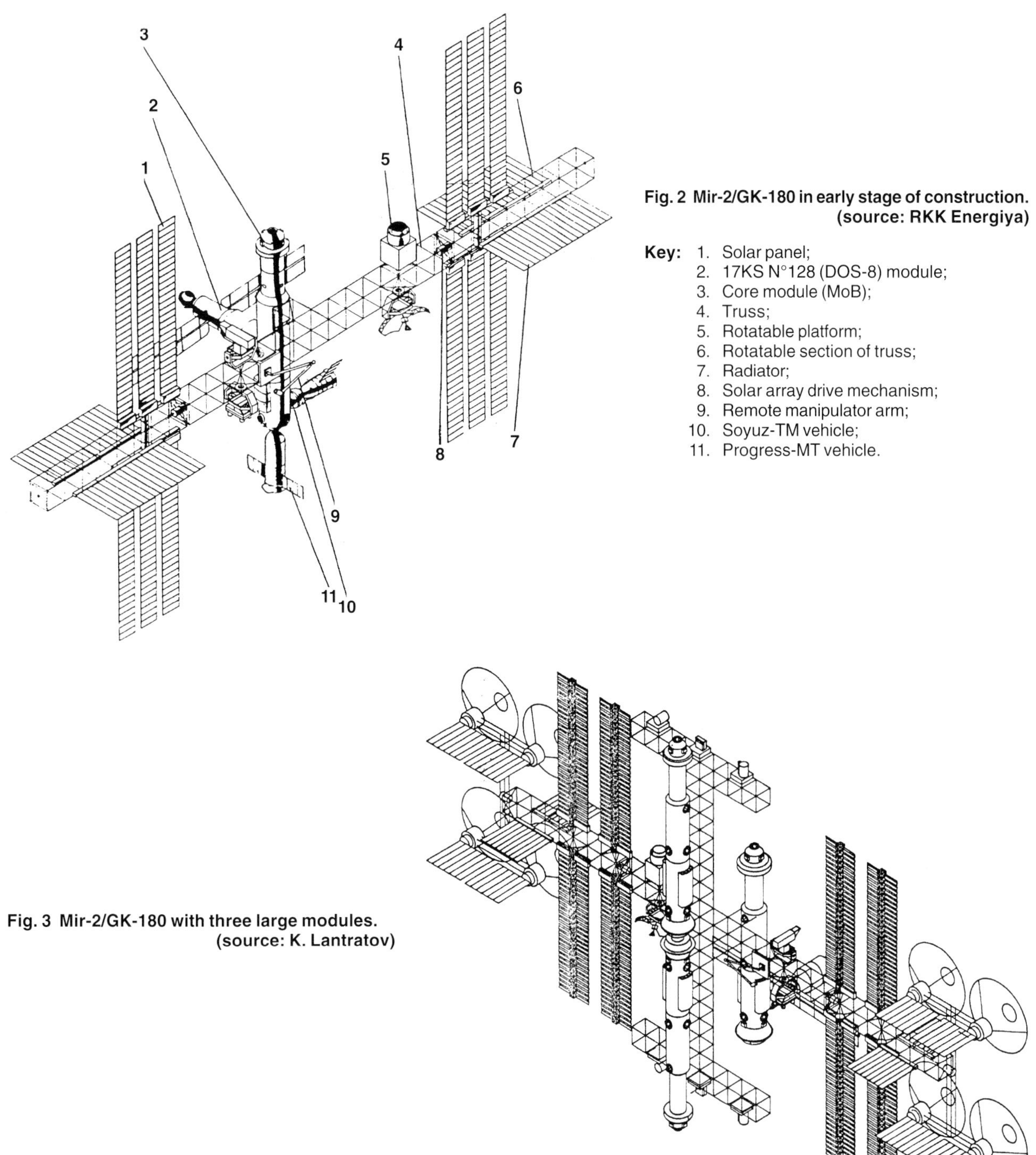

**Fig. 2 Mir-2/GK-180 in early stage of construction.
(source: RKK Energiya)**

Key: 1. Solar panel;
2. 17KS N°128 (DOS-8) module;
3. Core module (MoB);
4. Truss;
5. Rotatable platform;
6. Rotatable section of truss;
7. Radiator;
8. Solar array drive mechanism;
9. Remote manipulator arm;
10. Soyuz-TM vehicle;
11. Progress-MT vehicle.

**Fig. 3 Mir-2/GK-180 with three large modules.
(source: K. Lantratov)**

to the standard 51.6° due to mass constraints. Servicing the giant station would be a variety of vehicles: Buran, a Zenit-launched cargo ship derived from Progress and a Zenit-launched reusable crew ferry called Zarya (14F70), essentially an enlarged Soyuz descent module covered with the same heat-resistant tiles as Buran. Alternatives for Zarya were the standard Soyuz-TM or an advanced Soyuz-TM launched by Zenit.

It was not until 25 December 1989 that the plans for Mir-2/180GK were officially sanctioned by a State Commission of the USSR Council of Ministers. By now DOS-8 was expected to go up in 1994 and would receive the giant Energiya-launched module in 1995. In 1990 blueprints were made for the changes that were required to DOS-8 in its role as the living quarters for 180GK. Instead of six passive docking ports of the 'probe/drogue' type, the module would have four androgynous docking ports, two axial and two lateral. The other two lateral docking ports originally foreseen for DOS-8 were no longer needed. One would be turned into a 1 metre diameter EVA exit hatch and the other would simply be covered with a plate. Other planned changes included a new satellite communications system ('Kvant-OK') and modified Kurs rendezvous antennas.

Statements by Soviet officials in 1989-90 indicated that the launch date for Mir-2/180GK kept slipping further and further into the 1990s and financial cutbacks eventually forced these ambitious plans to be scrapped, the first report of cancellation coming in April 1991 [3]. It was necessary to return to the old plans for a more modest 20-tonne core module with add-on modules.

Back to Basics

"Mir 1.5"

One option considered was light-heartedly referred to by some as "Mir 1.5" and called for swapping DOS-7 with DOS-8. In fact, this scenario had already been foreseen in the Mir-2/180GK Draft Plan and, if implemented, would have meant that the giant station would be built without DOS-8. According to Soviet media reports in mid-1991 the idea was to orbit DOS-8 with a Proton rocket and then tow it to Mir with the help of Buran, following which one or more add-on modules would have been transferred to the new core module. The Mir Base Block and the obsolete modules would subsequently have been deorbited with the help of Progress cargo ships [4]. A more elaborate version of this plan was presented by NPO Energiya general designer Yuriy Semyonov at the IAF Congress in Montreal in the autumn of 1991 and envisaged launching DOS-8 in 1994 and keeping it attached to Mir for about two years. During this period Buran would have delivered a prototype biotechnology module called 37KBT. After the transfer of the Priroda module to DOS-8, Mir would then have been discarded, paving the way for the four-year assembly of the Mir-2 complex. During that time Buran would have been used to regularly deliver and return to Earth two operational biotechnology modules known as 37KBT N° 1 and N° 2. A large truss structure with solar panels would have been added, with assembly expected to be complete by 2000 [5].

The 1992 Mir-2 Design

Eventually, the choice fell on another option that had been considered parallel to the "Mir 1.5" plan, namely to launch DOS-8 after DOS-7 had outlived its usefulness and expand it with add-on modules. This would also allow the new station to be placed into a 65° inclination orbit, as had been the intention all along. Launches to it were to be staged from both Baykonur and Plesetsk. Detailed plans for the new Mir-2 station began appearing after the middle of 1992 [6]. On 24 November 1992 they were approved by the Council of Chief Designers, a body including representatives from the various design bureaus involved in the station. Mounted on top of the Base Block would have been a large cross beam known as the Science Power Platform (Russian abbreviation NEP). The NEP was to be outfitted with retractable solar panels similar to those used on Mir's Kristall module and also with solar parabolic concentrators, known in Russian terminology as a 'solar gas turbine installation' and in NASA jargon as a 'solar dynamic power system'. Unlike a photovoltaic system, such concentrators utilise the Sun's heat instead of its light for the production of power. The heat is collected in a receiver which is located near the focal point of a parabolic mirror and power is generated exactly the same way as in a power station on Earth, namely by heating a fluid which in turn rotates a turbine. Since a heat/gas-driven turbine is a much more efficient power converter than a solar cell, the mirror has to be only one fourth the area of a solar array to generate the same amount of power. Other devices to be mounted on the NEP were radiator panels, small engine units (similar to the VDU unit on Mir's Sofora mast) and small rotatable science platforms comparable to the ones on Kvant-2 and the Vega Venus/Halley probes. The ball-shaped multiple docking adapter of DOS-8 was once again to be outfitted with four lateral docking ports as in the original design. All six docking ports were to be of the 'probe-drogue' type.

The add-on modules for DOS-8 represented a radical departure from the bulky 20-ton modules of Mir. The Mir modules themselves had gone through a long evolutionary path ever since the approval of multimodular space stations in 1976. The original idea was to launch small Soyuz-based specialised modules that could regularly be changed out. In 1979, after the design bureau in Fili joined the Mir project, that idea was dropped in favour of the Proton-launched 37KS modules, which consisted of a laboratory section and a heavy space tug derived from the FGB units of the Transport Supply Ships (TKS), originally developed to ferry crews and cargo to Almaz. One of the 37KS modules, Kvant, docked with Mir in 1987, but the others were cancelled in 1983/84 in favour of the 20-ton 77KS modules that eventually flew to Mir as Kvant-2, Kristall, Spektr and Priroda. These were also designed by the Fili bureau (by now called KB Salyut) and based on the FGB.

The proposals tabled for Mir-2 in 1992/93 marked a return to some of the cancelled module concepts for Mir and also eliminated any role for KB Salyut, which in 1993 became part of the Khrunichev Centre. First, they would be much lighter than the 77KS modules, allowing them to be launched by Soyuz or Zenit rockets. Second, they would be ferried to the station by detachable space tugs, no heavy derivatives of the FGB as in the case of the 37KS modules, but units derived from the propulsion compartment of the Progress-M cargo ship. The four radial docking ports of Mir-2 were to be occupied by a Docking Compartment, a Service Module, a Technological Module and a Biotechnological Module. An alternative plan to launch the latter two with Buran was given up fairly soon as funds for the Russian shuttle dried up after the collapse of the USSR (Fig. 4).

The Docking Compartment had an airlock for spacewalks and an aft androgynous APAS docking port mainly intended for Buran dockings, which would have required it to be temporarily moved to the front axial port of the Base Block. The main functions of the Service Module (not to be confused with the identically named corner-stone of the ISS Russian segment) were to provide attitude control through the use of control moment gyroscopes ('gyrodins' in Russian terminology), to store and distribute power provided by the solar panels and solar parabolic concentrators and to make sure that excess heat from the various modules was transported to the radiator panels on the NEP and the Base Block. It also had an aft docking port to receive Soyuz-TM and Progress-M vehicles. No repositioning of the module was required for such dockings. The Technological and Biotechnological Modules would have been used for the production of pharmaceutical products as well as pure crystals and alloys.

The Docking Compartment weighed about 3-4 tonnes and together with the attached Progress-M propulsion compartment remained within the payload limits of the Soyuz-U rocket. The Service Module and the two production modules were based on a new resupply ship designed for launch by the Zenit booster. An initial version of this Zenit-launched resupply ship, essentially an enlarged Progress vehicle, had been conceived around 1988 and was supposed to fly to the giant 180GK/Mir-2 space station. Known as Progress-MT or 11F615A75 (the standard Progress-M being 11F615A55), it would have had enlarged cargo and propellant compartments [7] (Fig. 5). A slightly modified version for the new Mir-2 appeared on the drawing boards in the early 1990s. Its index was 11F615A77 and it later became known as Progress-M2 (not to be confused with the second vehicle in the Progress-M series) (Fig. 6). The overal mass of the ship was 13.3 tonnes and it was 12.6 metres long. It had a 5.3-tonne compartment combining the functions of the Progress-M propulsion and propellant compartments, the propellant being used to either refuel the station or provide additional fuel to the ship's own engines. Attached to this section would be an 8-tonne cargo compartment (2.3 tonnes empty with a maximum of 5.7 tonnes of cargo) with a docking port. The plan apparently was for Progress-M and Progress-M2 to be used simultaneously to resupply Mir-2.

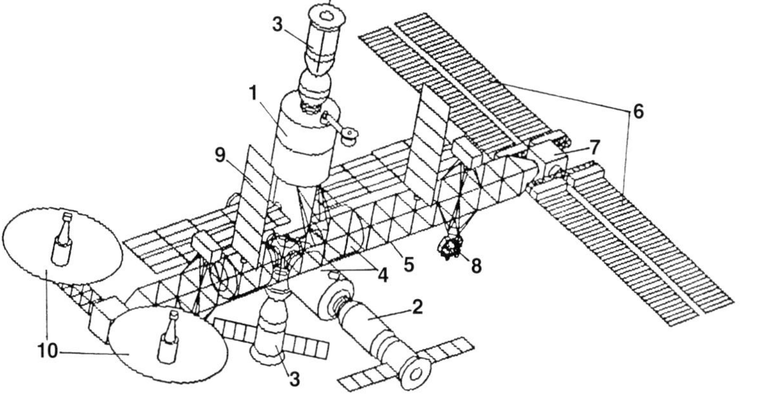

Fig. 4 Mir-2 configuration in 1992.
(source: K. Lantratov)

Key:
1. Base Block (DOS-8);
2. Progress-M2 cargo ship;
3. Soyuz-TM vehicle;
4. Specialised modules;
5. Truss;
6. Solar arrays;
7. Solar array drive mechanism;
8. VDU thruster package;
9. Radiator;
10. Solar parabolic concentrators.

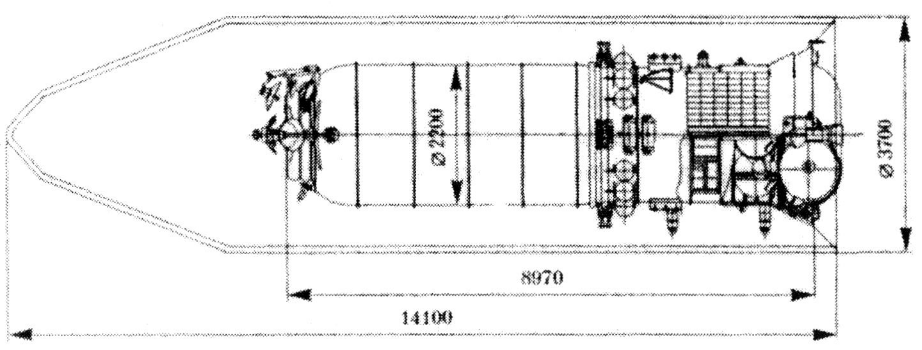

Fig. 5 A version of Progress-MT briefly considered in 1999 for launch by the Yamal rocket.
(source: RKK Energiya)

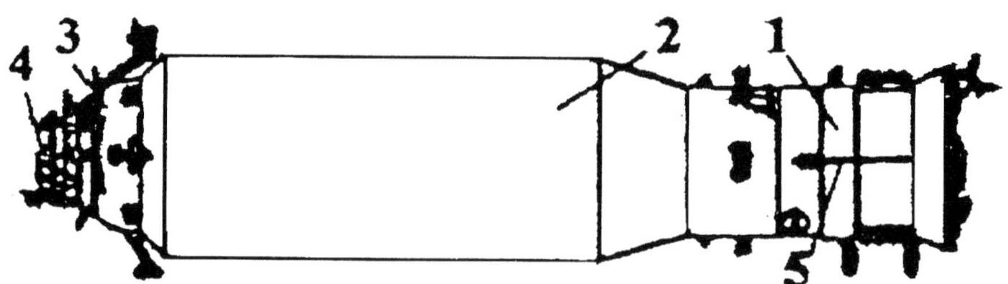

Fig. 6 The Progress-M2 cargo ship.
           (source: Novosti Kosmonavtiki)

Key: 1. Propulsion and propellant
            compartment;
         2. Cargo compartment;
         3. Forward section;
         4. Docking port;
         5. Solar panel.

Progress-M2 could easily be reconfigured as a permanent add-on module by changing the 8-tonne compartment into a laboratory. Two other Progress-M2 derivatives were later described for the International Space Station and it is likely that they were originally designed for Mir-2 as well. One was a tanker version in which the 8-tonne cargo compartment would have been replaced by an unpressurised propellant section. This vehicle could also act as a tug for larger modules and payloads. The other was an assembly version that would substitute the 8-tonne compartment for a large structure to be used in station assembly. The module and assembly versions would have discarded their propulsion compartments after reaching the station [8].

According to the 1992 plans construction of the Mir-2 complex was to begin with the launch of the DOS-8 core module in the first quarter of 1996, followed shortly afterwards by the assembly of the NEP truss using several Progress or Buran launches. That same year DOS-8 was to be joined by the Docking and Service Modules, with the Biotechnological Module going up in 1997 and the Technological Module in 1998. The station was expected to remain operational for at least ten years and would be permanently occupied by two to three cosmonauts.

The 1993 Mir-2 Design

Sometime in the first half of 1993 it was decided to expand Mir-2 with two so-called Universal Docking Modules (Russian abbreviation: USM) that provided extra docking opportunities for modules and transport ships and eliminated the need for lateral docking ports on the DOS-8 docking adapter. The USMs were to be launched by Zenit and delivered to the station by a detachable Progress-M2 propulsion compartment. They had about the same length and mass as the other Progress-M2 based modules, but consisted of three sections rather than one long cylindrical compartment. The front section was a small cylinder with a single axial docking port, the middle section was about the same diameter as the Progress-M2 cargo section and the aft section was a square-shaped docking adapter with five docking ports, four lateral and one axial. Most or all of the docking ports were supposed to be of the androgynous APAS type. In the previous Mir-2 design only the Docking Compartment was supposed to have an APAS port (Fig. 7).

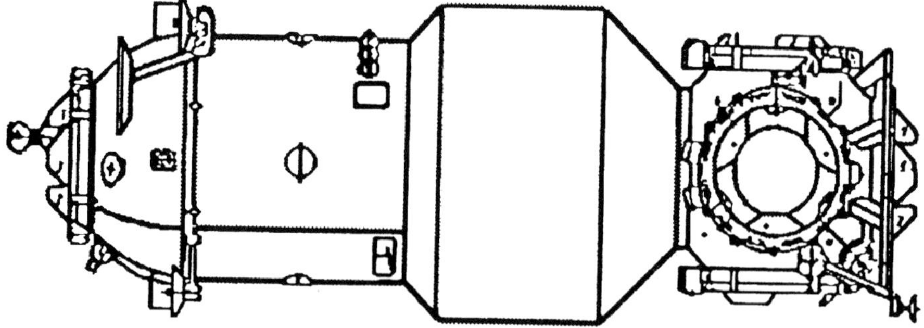

Fig. 7 The Universal Docking Module
with APAS docking ports.
           (source: Novosti Kosmonavtiki)

USM N°1 was to be attached to the front docking port of DOS-8 and USM N°2 to one of the lateral docking ports of USM N°1. The Biotechnology and Service Modules would be docked to the lateral ports of USM N°1 and the Technology Module plus a new Remote Sensing Module to the lateral ports of USM N°2. There would now be two Docking Compartments, one attached to USM N°1 and the other to the aft port of the Base Block. This gave cosmonauts more flexibility in perfoming spacewalks, making it possible to easily gain access to any part of the orbital complex. Another novelty was the installation of two pressurised sections on the NEP with gyrodins and storage batteries [9] (Fig. 8).

Given the new international climate Mir-2 would probably not have remained an all-Russian venture. In the summer of 1992 the European Space Agency began expressing interest in participating in the project, which would have given ESA a back-up option for pursuing its man-related space efforts in case the politically fragile Freedom project fell victim to

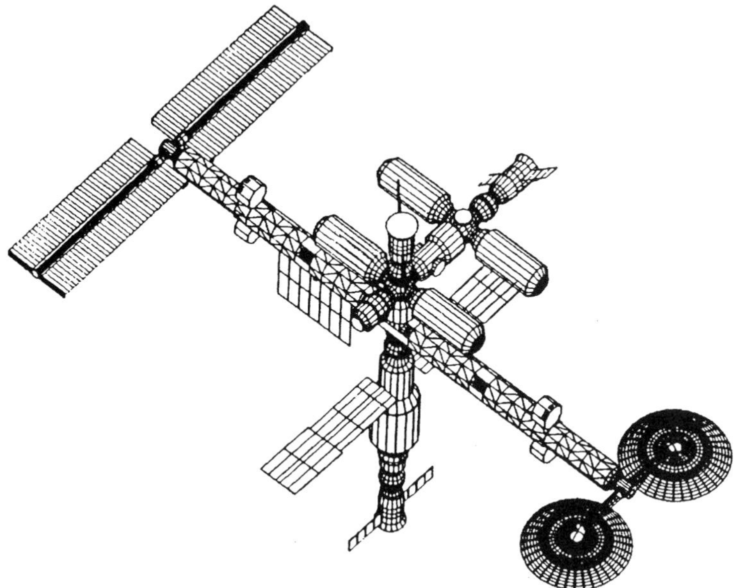

Fig. 8 Mir-2 with two Universal Docking Modules.
(source: RKK Energiya)

budget cuts. ESA showed particular interest in the 65° inclination of the Mir-2 complex. One early option studied was to fly the so-called Euro-Russian Technological Complex (ERTC) in conjunction with Mir-2. This was a European-Russian version of the Man-Tended Free Flyer (MTFF), a free-flying element of the Columbus programme that was no longer seen as a viable option for Freedom. It would have been delivered to the station either by a Progress-M or Progress-M2 propulsion compartment [10]. In November 1992 the ESA ministerial meeting in Granada gave the go-ahead for expanding Euro-Russian space co-operation, one aspect of which was to be joint work on the Mir-2 station. Co-operative ventures studied in the following months were the modernisation of Progress and Soyuz vehicles, a jointly built EVA spacesuit and a robotic arm called ERA (European Robotic Arm) [11]. Another potential partner was Japan, although no concrete proposals seem to have been worked out for Japanese contributions to Mir-2.

Merging Mir-2 with Freedom

Early Discussions (March-September 1993)

The first discussions on Russian participation in the Freedom project began in late 1991 and focused exclusively on the use of Soyuz-TM as an interim crew rescue vehicle for the joint US/European/Canadian/Japanese space station (see Soyuz section). US-Russian space co-operation got a further boost on 17 June 1992 when US President George Bush and Russian President Boris Yeltsin signed an agreement on joint manned flights during their first summit in Washington. This provided the basis for an agreement between NASA and the Russian Space Agency, signed in Moscow on 5 October 1992, in which the details of those flights were worked out. The plan was for a Russian cosmonaut to fly a mission aboard the US Space Shuttle in 1994 and for a US astronaut to spend about 90 days aboard the Mir space station in 1995, with the Space Shuttle performing a rendezvous and docking with Mir to bring back the US astronaut at the end of his mission.

The first move towards an actual merger of the Freedom and Mir-2 programmes seems to have come in early 1993. On 11 February 1993 NPO Energiya officials received an invitation from Boeing to discuss various aspects of US/Russian space co-operation. A delegation of NPO Energiya headed by general designer Yuriy P. Semyonov arrived at the Boeing headquarters in Seattle on 5 March 1993 for a week of joint discussions on 12 possible areas of co-operation. One of these was the creation of an international space station comprising elements of both Mir-2 and Freedom [12]. Russian Space Agency director Yuriy Koptev reportedly also discussed this option in a meeting with NASA Administrator Dan Goldin in Washington that same month [13].

On 15 March, just one day after the NPO Energiya team returned to Russia, Semyonov and Koptev sent a joint letter to Goldin outlining their proposals for such a station. It would orbit at an inclination higher than 50°, a compromise between the planned 28° inclination of Freedom and the 65° inclination of Mir-2. They stressed that the merger of the two space stations would be cost-effective for both sides. Under the Semyonov/Koptev proposal the Russian segment would consist of four Mir-2 elements: the DOS-8 Base Block, a Universal Docking Module, a

Docking Compartment with airlock and the Service Module. The Universal Docking Module would act as the interface between the Russian and US segments. Soyuz-TM and Progress-M vehicles would be used to ferry up crews and cargo. Station assembly was to begin with three Russian launches (DOS-8, the Universal Docking Module and the Docking Compartment) after which a Canadian robot arm would be deployed to assist in further assembly activities. These would include the addition of the US Lab, the US Hab and the European and Japanese modules. A total of fourteen assembly flights would be needed to complete the station. The station would be able to house a permanent three-man crew by the sixth flight in 1997 and would be continuously manned by nine astronauts in 2000. By contrast, the Freedom plans called for *beginning* permanent occupation in 2000 [14].

The Russian proposal came at a time when the Freedom project was running into ever deeper trouble. Originally approved by President Reagan in 1984, it was seriously over budget and still several years away from its first launch. On 18 February, less than a month after taking office, President Bill Clinton had ordered yet another redesign of Freedom and directed its cost to be reduced by more than half and to advance the construction schedule. On 9 March NASA Administrator Dan Goldin set up a redesign team that was to present various options for the redesign to Clinton by early June.

Meanwhile, officials from the White House, NASA and the Defence, Commerce and State Departments had begun debating the possibility of further involving the Russians in the station effort, possibly in response to the Koptev/Semyonov letter. Both Clinton and Vice-President Al Gore reacted favourably to the idea when they were briefed about it on 1 April. During a summit between Clinton and Yeltsin in Vancouver on 3-4 April the Russians were officially invited to offer advice on the station redesign effort. Another outcome of the summit was the formation of a high-level commission led by Gore and Russian Prime Minister Viktor Chernomyrdin to chart the course of US-Russian co-operation in energy and space. Officially called the Russian-American Commission on Economic and Technological Co-operation, it became informally known as the Gore-Chernomyrdin commission [15].

From 22 April to 5 May a Russian delegation including Koptev, Semyonov, Polukhin (KB Salyut), Utkin (TsNIIMash) and Grigoryev (IMBP) was in the United States to discuss the use of Russian experience in the space station redesign with NASA, White House and Defence Department officials. Members of the delegation proposed two variants of co-operation. One would see the joining of Freedom and Mir-2 into a single station pretty much in the same way that Koptev and Semyonov had outlined in their 15 March letter to Goldin. Another possibility would be separate stations that would use the same transportation and rescue systems, which would still require them to be launched into the same inclination. The Russian officials stressed that they were nearing a decision on full-scale Mir-2 production and therefore needed to know soon whether they would be invited to take part in the international station. However, it soon dawned upon them that they had come to the US with too high expectations. NASA officials made clear that a combined station was beyond the scope of what the redesign team was studying. Instead, they consulted the Russians on the possibility of using the Proton rocket to launch large elements of the station (such as the European and Japanese modules). The Russians were also invited to evaluate the cost of building the primary American modules and their life support systems, with final outfitting taking place in the US. The delegation was confronted with widely differing opinions on the level of Russian involvement in the station and left the US with mixed feelings on the prospects of co-operation [16].

Faced with an early June deadline to present its report to the White House, the station redesign team simply did not have the time to integrate the radical Russian proposals into the options it had singled out by this time. Shortly after its formation in March the team had been bombarded with ideas from all sides. At least one of these had been for Freedom to be assembled in a 51° inclination orbit so that it could share rescue and supply vehicles as well as other components with Mir and/or Mir-2, allowing people and cargo to be ferried between the two stations [17]. While the team kept the 51° inclination orbit open as an option, the three redesign plans that emerged from their studies by late April limited Russian co-operation to the use of the Soyuz-TM lifeboat and Russian docking systems. At that point the team stopped accepting major changes to the three options so that NASA analysts could examine the price tag of each before the deadline expired [18].

Option A (nicknamed 'austere') was a scaled-down version of Freedom with the possible inclusion of a Lockheed propulsion unit ('Bus-1'), probably used on the advanced KH-11 spy satellites. Option B (nicknamed 'baseline') represented a minimal departure from the Freedom design and Option C (nicknamed 'the can') called for a single-launch core station that could later be expanded with additional modules. Clinton eventually picked a combination of Options A and B, a decision announced by the White House on 17 June. He also called for sweeping management and contract reforms and keeping spending under $11 billion over the following five years, a $4 billion reduction from the original estimates.

Important questions such as the station's inclination and Russian involvement remained unsettled and many Congress members continued to have second thoughts about the still impressive cost. Later that month an amendment aimed at shutting down the project was rejected in the House of Representatives with the narrowest possible margin (216-215).

NASA's redesign team, renamed the transition team, was now tasked with working out the technical and managerial details of the revamped station before 7 September. As the team got down to work, one of the major political stumbling blocks towards expanding Russian co-operation on the station was overcome. After intense negotiations in Washington between US and Russian officials in mid-July Moscow agreed to abide by the Missile Technology Control Regime, an international accord that limited sales and the transfer of rocket parts and technology. In May 1992 the US government had accused Russia of violating this agreement by selling cryogenic upper stages to India. In letters to Yeltsin and Chernomyrdin in mid-June 1993 Clinton and Gore had linked space station and commercial launch co-operation with Russia's acceptance of anti-proliferation guidelines. This drew an angry response from the Russians and reportedly led to the postponement of the first meeting of the Gore-Chernomyrdin commission, originally scheduled to take place in Washington in mid-June. The agreement, signed by Koptev and US Undersecretary of State Lynn Davis on 15 July, allowed Russia to sell the stages to India, but stipulated that controls be applied on the amount of technology transferred with the hardware. With this hurdle out of the way, the stage was set for the signing of two more key agreements the following two days. On 16 July Koptev and assistant US Trade Representative Peter Allgeier approved an accord to allow Russia into the commercial launch market and on 17 July Koptev and Goldin agreed to study the feasibility of further co-operation in the Shuttle/Mir and space station programmes by 31 August. The agreements would have to lead to a major new US-Russian co-operation pact to be signed by Gore and Chernomyrdin in Washington in early September [19].

On 27 July Goldin met with the chiefs of the European, Japanese and Canadian space agencies in an attempt to allay their fears about increased Russian co-operation in the space station. At the same time, NPO Energiya manager Yuriy Semyonov, appearing at a symposium in Virginia, voiced the hope that the resolution of the export disagreement would open the way for the combination of Russian and international station efforts. Two days later a team of Russian aerospace industry technicians and government representatives began arriving in the US for a month of brainstorming with the station transition team under the agreement reached between Koptev and Goldin on 17 July. The 26-member team included representatives from NPO Energiya, the Khrunichev Centre, TsNIIMash, Star City, IMBP, the Scientific Research Institute of Thermal Processes (NIITP) and the Russian Space Agency. Although a complete merger of Freedom and Mir-2 was supposed to be only one of the options to be discussed, it evolved into an ever more attractive idea as the weeks progressed. Not only would it lead to major cost savings, it was also felt that it would lessen the temptation for some Russian companies to violate international missile proliferation guidelines.

By the end of the month the Russian team produced a 137-page report in which it concluded that Russia could provide about $2.5 billion worth of services and hardware to the station and save the US an estimated $7 billion in expenditures on the programme [20]. On 26 August Goldin and Koptev once again discussed the ambitious plans for co-operation and reportedly agreed to work together on an international space station that would supersede the revamped programme approved by Clinton in June [21]. Finally, on 2 September Gore and Chernomyrdin signed a space pact in Washington calling for a phased station effort. Phase One would see the expansion of Shuttle/Mir operations, with Mir being available for up to two years of aggregate US astronaut time on board. Phase Two would be aimed at 'the use of a next-generation Russian Mir base block together with a laboratory module and the Shuttle' to provide an interim human-tended science capability. According to the agreement, 'successful implementation of this phase could constitute a key element of a truly international space station'. The pact called for working out a detailed plan for such an international station before 1 November [22].

As had been the case in June, the transition team did not have enough time to include the Russian proposals into the revised station plan it was to present to President Clinton by 7 September. What Clinton got on his desk instead was a 37-page plan for a station known as 'Alpha', based largely on the Option A design produced during the redesign work in the spring. A key feature of the station was that the basic configuration remained unchanged regardless of how much Russian hardware would be used. It would maintain the same arrangement of the US, ESA and Japanese modules, the Canadian manipulator arm, the truss and the solar arrays and radiators. That basic configuration could then be complemented in three different ways. In the first two, either the Lockheed Bus-1 or Russian FGB space tugs could be added to provide propulsion, guidance, navigation and control (Fig. 9). In the third option the basic Alpha could be combined with Mir-2 hardware, with the Russians becoming full-fledged partners in the station [23].

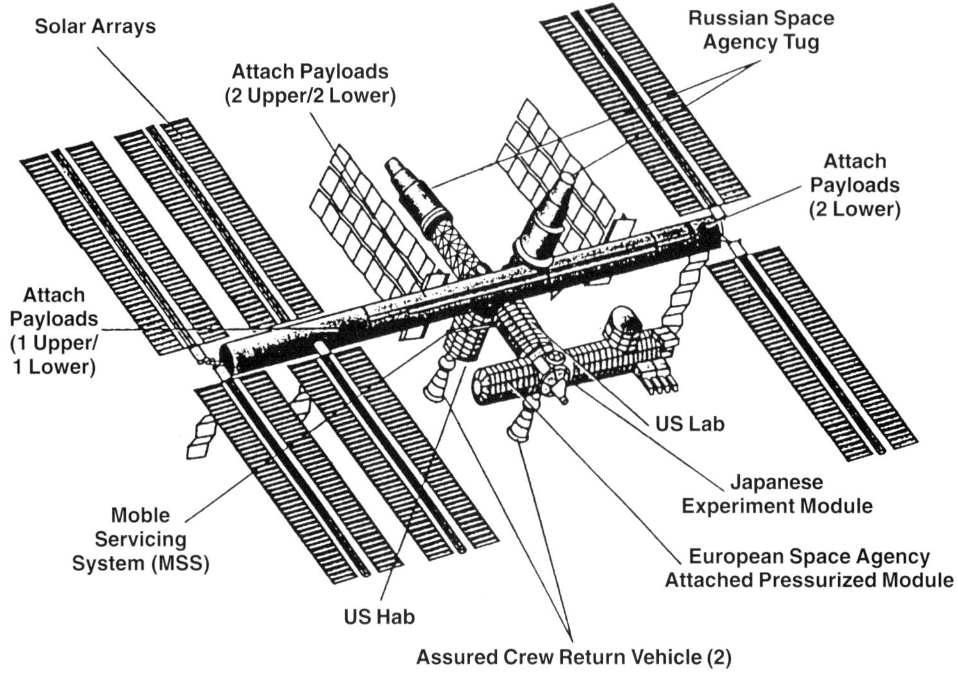

Fig. 9 Alpha concept with two FGB tugs.   (source: NASA)

Reaching a Final Concept (September 1993-September 1994)

In the plan for the joint station presented by the Russians on 26 August 1993 the Russian segment was to consist of the following Mir-2 elements (Fig. 10):

- the Base Block
- three Universal Docking Modules
- the Docking Compartment (with airlock)
- the Service Module
- three Research Modules (a technological, a biotechnological and a remote sensing module)
- a downsized version of the Science Power Platform with two solar parabolic concentrators

Notable in the plan was the heavy reliance on Russian elements in the early stages of construction and the ability to have a permanent crew on board from the very start. The first five launches would all be Russian. Assembly was to begin in late 1996 with the launch of the DOS-8/Mir-2 Base Block, which was to be boarded immediately by a three-man resident crew ferried up by Soyuz. Subsequently, two Universal Docking Modules were to be docked in tandem to DOS-8 to permit further expansion of the Russian segment and to provide an interface with the US segment. In the next step the Docking Compartment was to be hooked up with UDM Nr. 2. It was to act as the berthing place for the first US Space Shuttle assembly mission, which was to attach the US Node-1 to UDM Nr. 2, setting the stage for the construction of the 'Western' side of the station. The zenith port of UDM Nr. 1 would later be occupied by the Science Power Platform, which required two Russian launches, and the nadir port was to receive UDM Nr. 3. This in turn was to serve as the core for the final four Russian modules (the Service Module and the three Research Modules). Assembly of the station was to be finished by late 2000 [24].

Now that the Gore-Chernomyrdin commission had given approval to come up with a firm design for the station, Russian and US experts got down to working out the details. An idea that soon emerged was to replace the two Universal Docking Modules in the longitudonal axis of the station by a single FGB tug. It seems that the NASA transition team had already been informed about the FGB's capabilities during the August negotiations with the Russians and it was seen as a potential alternative to the Lockheed Bus-1 in the Alpha design even if the Russians were not to become full partners. Not only was FGB expected to be cheaper, unlike Bus-1 it could be refuelled in orbit. In late September a delegation of NASA and US industry officials was given more information about the FGB during a visit to the Khrunichev Centre in Moscow. On 4 October a Khrunichev representative presented a plan for integrating the FGB into the international space station during a meeting with NASA, Boeing and NPO Energiya

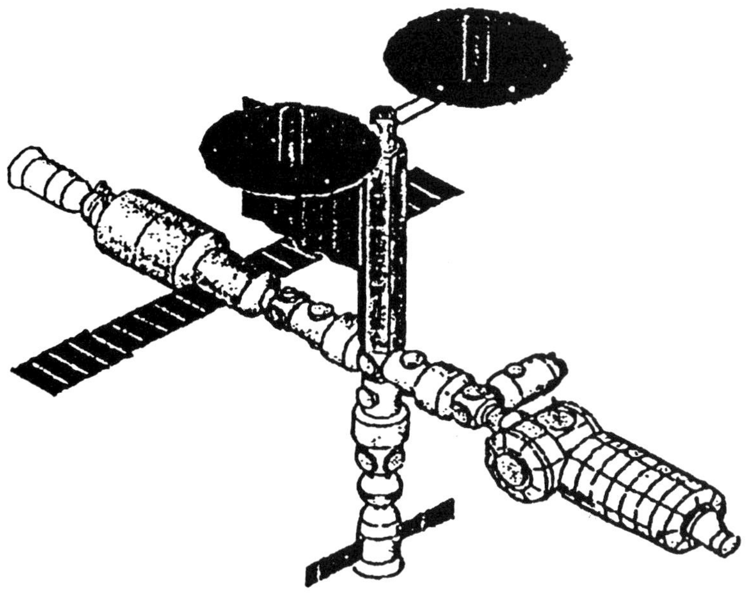

Fig. 10 ISS configuration with three Universal Docking Modules proposed by Russia in August 1993.
(source: Novosti Kosmonavtiki)

officials at the Russian Space Agency headquarters in Moscow. The plan was accepted that same month. FGB was to become the first building block of the station and would take care of guidance, navigation, control and station reboosting during the early stages of assembly. Since the FGB had not been planned to be built as part of Mir-2, NASA agreed to pay for the construction of the vehicle.

From an engineering standpoint, there was no compelling need to include the FGB into the ISS. As DOS-8 was supposed to take over most of the FGB's critical functions just months after the latter's launch, station assembly might just as well have been kicked off with the DOS-8 launch as the Russians had proposed in the first place. Clearly, politics played an important role in this decision. With the US footing the bill for its construction, the first element of the station would now essentially become a joint Russian/American module and therefore somewhat alleviated the heavy emphasis on Russian launches in the early stages of assembly, something which the US side had clearly not been comfortable with. On the other hand, the choice of the FGB also solved an internal dispute within the Russian space industry itself. Ever since the early 1970s, NPO Energiya had delegated a significant portion of the design work on its DOS space stations to the Fili branch of Chelomey's bureau. When that bureau became part of Energiya as KB Salyut in 1981, it acquired an even more important role by taking on the design of Mir's add-on modules. However, after having separated from Energiya in 1988, KB Salyut was virtually sidelined for Mir-2, a situation which did not improve after it became part of the Khrunichev Centre in June 1993. Under the international space station assembly plan put forward in late August 1993, virtually all the Russian elements were to be both designed and built by NPO Energiya. Khrunichev's role was confined to that of a subcontractor to Energiya in building the former Mir-2 Base Block. Not satisfied with playing second fiddle to Energiya, Khrunichev Centre director Anatoliy Kiselyov reportedly urged his designers to come up with a plan to increase the centre's involvement in the station. The FGB, both designed and built by Khrunichev, perfectly fit the bill [25].

Another change made to the Russian segment in the autumn of 1993 was to delete the two solar parabolic concentrators from the Science Power Platform. According to the original Russian proposals these were to have been mounted in the early stages of assembly, but NASA engineers doubted they could be developed in such a short time. They also feared it would be difficult to dissipate the heat produced by the concentrators' turbines, something which could only be achieved with huge radiators. A NASA proposal to decrease the capacity of each concentrator from 10 kilowatts to 2-3 kilowatts was turned down by the Russians on the grounds that the whole idea of the concentrators was to produce large amounts of energy. The final compromise was to indefinitely postpone the installation of the 10 kWt concentrators and deploy them at a later stage on the US truss rather than the Science Power Platform. Instead, the Science Power Platform would be equipped with standard solar arrays producing roughly the same amount of power as the solar concentrators. There were plans to test the technology for the solar parabolic concentrators in the Shuttle-Mir programme, but eventually the concentrators were dropped altogether from the ISS design [26].

The resulting 'Addendum to the Space Station Alpha Programme Implementation Plan' was officially approved by

Goldin and Koptev on 1 November and presented to the White House the same day. The details were outlined at a press conference by Goldin and Koptev three days later. The Russian segment (also known as 27KSM) now consisted of the following elements:

- the FGB Energy Block
- the Service Module (NASA and Russian acronyms: SM)
- the Docking Compartment (NASA acronym: DC, Russian acronym: SO)
- one Universal Docking Module (NASA acronym: UDM, Russian acronym: USM)
- the Life Support Module (NASA acronym: LSM, Russian acronym: MZhO)
- three Research Modules (NASA acronym: RM, Russian acronym: IM)
- the Science Power Platform (NASA acronym: SPP, Russian acronym: NEP)

Rather confusingly, the former Mir-2 Base Block was now renamed the Service Module, the same name that had been used for one of the Mir-2 add-on modules. What used to be the Mir-2 Service Module was now replaced by the Life Support Module, a module with additional life support systems for the Russian segment. Assembly was now slated to begin in May 1997 with the launch of the FGB Energy Block, followed one month later by the Docking Compartment. In July 1997 NASA was to add the Node-1, with the Russians putting up the Service Module later that same month. The Science Power Platform, requiring one US and one Russian launch, was now to be mounted on the zenith port of the FGB, with the zenith and nadir ports of the Service Module to be occupied by the Life Support Module and the single remaining Universal Docking Module respectively. Assembly of the whole station was expected to be finished in October 2001 [27] (Fig. 11).

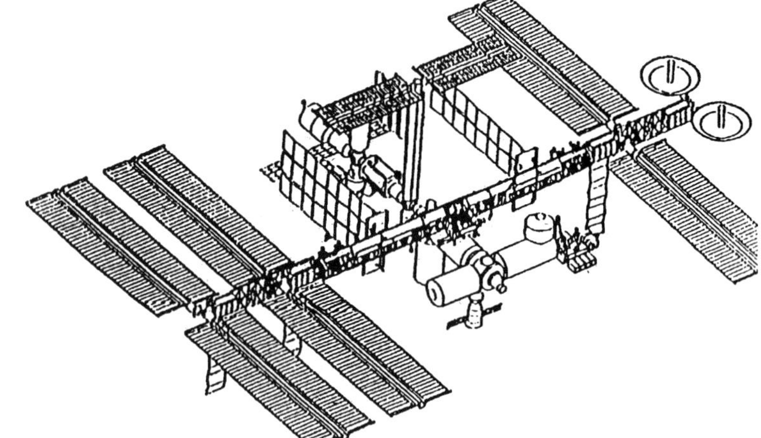

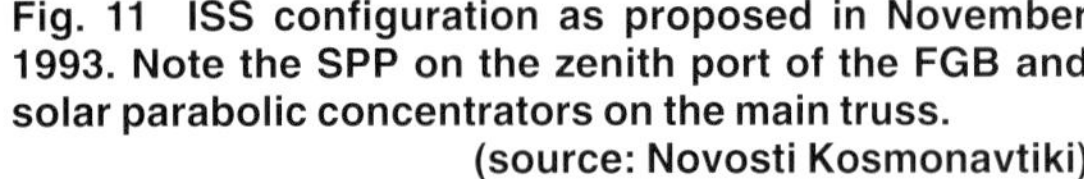

Fig. 11 ISS configuration as proposed in November 1993. Note the SPP on the zenith port of the FGB and solar parabolic concentrators on the main truss.
(source: Novosti Kosmonavtiki)

One more change in the Russian segment took place before the end of 1993. The Science Power Platform was moved from the FGB to the Service Module's zenith docking port and all launches needed to assemble it would now be performed by the Russians themselves [28]. The Life Support Module, originally supposed to occupy the Service Module's zenith port, was transferred to the FGB nadir port, with the Docking Compartment moving from the FGB nadir to the FGB zenith port.

In the weeks after the 4 November announcement two major political hurdles were overcome that cleared the path for full Russian participation in the station. On 29 November the White House won backing for the plan from key congressional leaders. Many within Congress had expressed second thoughts about Russia's involvement, especially after more political upheaval in Moscow in the preceding weeks. On 6 December representatives from 10 European countries, Japan, Canada and the US formally invited the Russians to take part in the space station project. Finally, on 16 December the Russians officially accepted the invitation during the second meeting of the Gore/Chernomyrdin commission in Moscow, completing almost nine months of lobbying. It took several more months of intense negotiations to work out the organisational and financial details of the joint undertaking. On 23 June 1994, in a ceremony timed to coincide with the third meeting of the Gore/Chernomyrdin commission in Washington, the heads of the US and Russian space agencies signed a $400 million contract under which NASA would pay $305 million to Russia for joint Shuttle/Mir operations and $95 million for Russian contributions to the International Space Station. On the organisational side, the Russian Space Agency would be responsible for co-ordinating work on the ISS Russian segment and carrying out negotiations with foreign partners, while RKK Energiya would act as prime contractor.

On 28 September 1994 NASA released a new assembly sequence, which aside from a slip in the FGB's launch from May to November 1997 showed some important changes in the Russian segment. Some of these resulted from a 10 August meeting where RKK Energiya general designer Yuriy Semyonov had signed the Draft Plan for the Russian segment. Most notable was the addition of the so-called Docking and Stowage Module (NASA acronym: DSM, Russian acronym: MSS), which was to provide extra storage room and had an aft docking port for receiving Russian ferry vehicles as well as the planned US lifeboat. The DSM was to be docked to the FGB nadir port, necessitating another relocation of the Life Support Module, which was now moved to one of the lateral docking ports of the Universal Docking Module.

The new schedule also reflected NASA's attempts to ease Russia out of the 'critical path' as much as possible. In the original assembly schedule there was a heavy reliance on the Russian Science Power Platform to provide energy for the US segment in the early stages of station construction. The huge US solar arrays could not be installed until after the deployment of the truss on the US Lab. In July 1994 NASA decided to push forward the launch date of one of its solar arrays from August 2000 to September 1998 by temporarily mounting it on on top of the US Node 1, mainly to satisfy the energy requirements of the US Lab. At that time a preliminary decision was made to delay the launch of the SPP to October 2000. However, in the September 1994 assembly schedule that date was moved *back* to late 1998/early 1999 in order to avoid the opposite situation, namely that the Russian segment would have to draw too much power from the US solar arrays [29].

Another pivotal Russian element in the early stages of station construction was the Docking Compartment, which was to play a crucial role in connecting the Russian and US segments. The plan was for the DC to link up with one of the FGB's longitudonal docking ports and subsequently receive the Shuttle on the Node-1 delivery mission. Once the Shuttle had arrived, the FGB was to undock, reposition itself so that one of its radial ports faced the DC/Shuttle combination and then redock. Next the Shuttle's remote manipulator arm would lift Node-1 out of the cargo bay and attach it to the now vacant 'front' port of the FGB (Fig. 12). Apparently, NASA considered this procedure too risky and opted instead for a mission in which the Shuttle's arm would capture the FGB and then directly dock it to the Node-1 still sitting in the cargo bay. This obviated the need for an early launch of the DC and no longer required it to be attached to the FGB zenith port, which was deleted from the FGB design. Instead, the Docking Compartment was transferred to the UDM, which now had all its four radial docking ports occupied (DC, LSM, RM-1 and RM-2). This meant that the LSM had to be fitted with an aft docking port in order to receive the third and final Research Module.

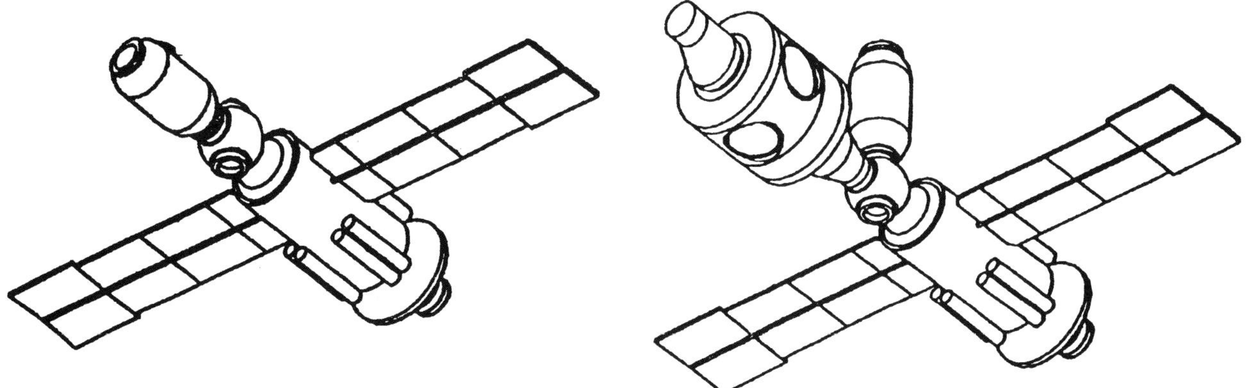

Fig. 12 Original location of the Docking Compartment to receive Node-1.　　　　　(source: Novosti Kosmonavtiki)

It was also at this point that the Russians decided to have three types of docking ports on their segment rather than one. The original plans for the Russian segment had called for the use of a single type of docking unit, namely the so-called APAS-89 (Russian short for Androgynous Peripheral Docking System) [30] (Fig. 13). A modified version of the Apollo-Soyuz APAS-75 system, it had originally been built for use on Buran's docking adapter and two of them had also been installed on Mir's Kristall module to receive Buran orbiters, small scientific modules and eventually also the US Space Shuttle. Reportedly, the problem with APAS was its relatively high mass, which would make Soyuz too heavy to transport a three-man crew to the station [31]. Therefore, the only APAS left on the Russian segment was on the side of the FGB docked to Node-1. All the other docking ports would be either of the standard 'probe-drogue' type used on Mir, officially called the Internal Transfer Docking System (Russian abbreviation SSVP) (Fig. 14), or a hybrid of the APAS and the SSVP

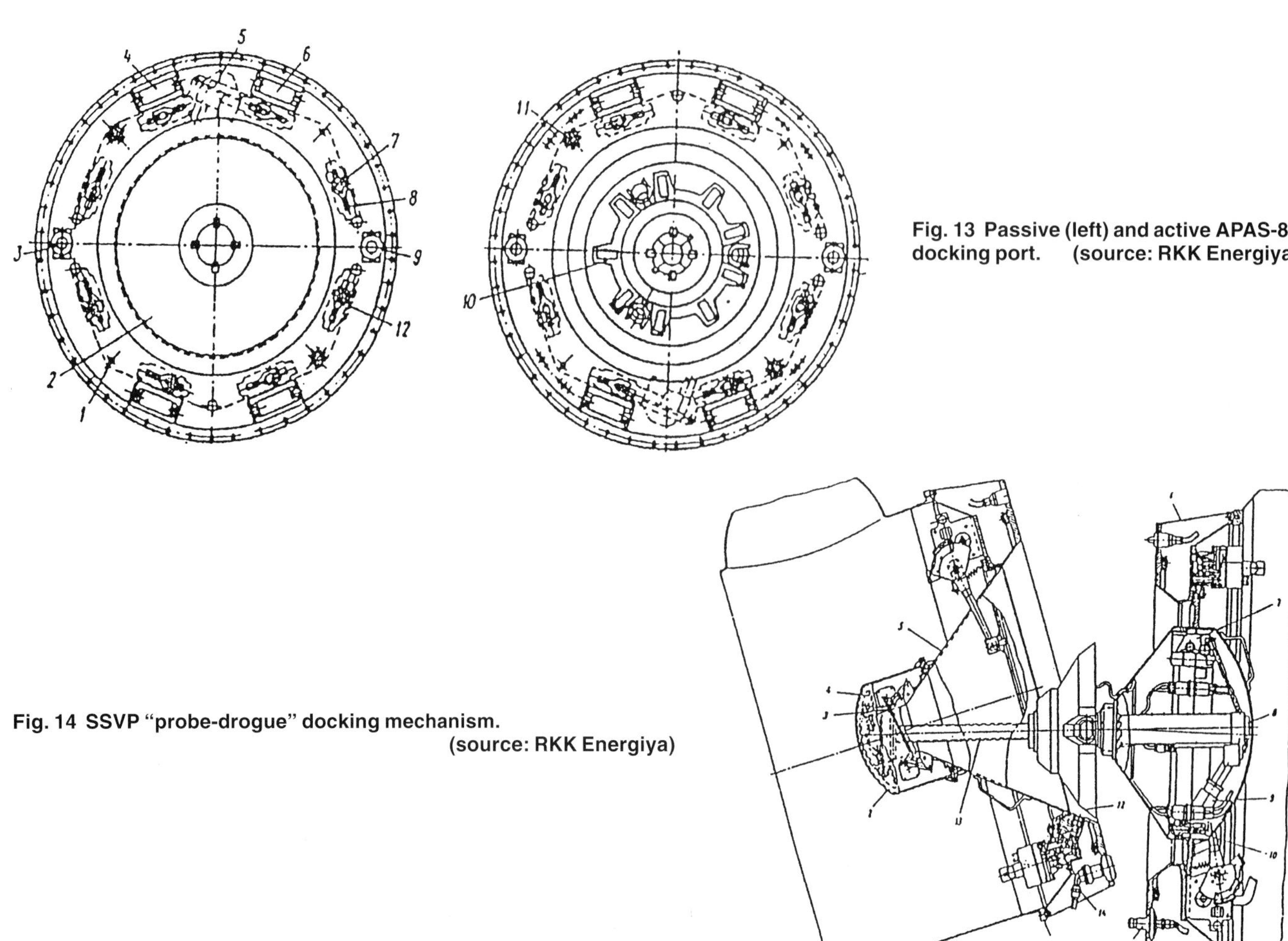

Fig. 13 Passive (left) and active APAS-89 docking port.   (source: RKK Energiya)

Fig. 14 SSVP "probe-drogue" docking mechanism.
(source: RKK Energiya)

called SSVP-M (Fig. 15). Providing a firmer connection than the SSVP, the SSVP-M had the probe-drogue linking mechanism of the SSVP and the periphery of the APAS. Since the SSVP and the SSVP-M were incompatible, every module and ferry was now restricted to a certain number of docking ports. Further complicating the docking situation was that all modules headed for the Universal Docking Module first had to link-up with its aft axial port and then engage one of two sockets on the UDM to be transferred to one of the four lateral ports (as had been the case on Mir's spherical docking adapter) [32].

The Mir Option (December 1995)

As things stood in 1995, Russia was to abandon its Mir complex in 1997 after seven Shuttle-Mir flights and then dedicate all its manned spaceflight resources to the assembly of the ISS. However, as time went on, it became clear that Mir would be able to continue in orbit for several more years, the more so because its two last add-on modules wouldn't be launched until the mid-1990s. All this faced the Russians with a dilemma. With Russia shouldering much of the responsibility for the early stages of station construction, the country could barely afford to operate two space stations simultaneously. In June 1995, only weeks after the launch of Spektr, a deputy of Yuriy Koptev raised the possibility of transferring Spektr and Priroda to ISS during a meeting with NASA station director Wilbur Trafton at the Paris Air Show. This would have involved the use of either the Shuttle or some sort of propulsion module, although the Russians would have had to incur the costs of the transfer.

In August officials of NPO Energiya and the Khrunichev Centre jointly began working out plans for making maximum use of Mir's resources in the build-up of ISS. These focused either on assembling the ISS in the vicinity of Mir or, more radically, on using Mir as a starting point for the assembly of the ISS. During a meeting at the Russian Space Agency on 13 October 1995 it was decided to present the plans to the Americans. In mid-December a Russian delegation visiting Houston outlined two scenarios in which the initial elements of the ISS would be attached to Mir, after which Mir elements would be gradually discarded as they became too old to remain in orbit. Few details of the Russian proposal were made public. Apparently, the plan was to first dock the FGB and Node-1 in orbit and then link up the combination with Mir's front docking port. The Service Module

Fig. 15 Passive SSVP-M hybrid docking port on the Service Module.
(source: NASA)

could then have been sent up to dock on the other side of Mir or could have stayed on the ground until the Mir core module had outserved its usefulness. Meanwhile, the US and the other partners would be free to build up their segment of the station on the Node-1 side. Under this plan the Russians would have been able to use the newly launched Mir modules to the maximum extent possible and could have scrapped the development of two of the three ISS Research Modules [33].

The plans, which had already drawn a lot of criticism from the US side in the weeks before the arrival of the Russian delegation, were rejected outright by the Americans. With the full backing of both Al Gore and Dan Goldin, two key House members in charge of NASA funding visited Moscow in early January to deliver the message that with such radical proposals the Russians were jeapordising Congressional support for the station and that, if necessary, the US and the other partners would go it alone. It was claimed that the Mir option would delay completion of the station by up to a year and would add billions of dollars to the station's price tag. For instance, it would force engineers to completely redesign electrical and mechanical connections between Mir and the Russian station components.

Later that month US and Russian negotiators hammered out a compromise plan that would allow Russia to keep Mir in orbit and still meet its space station obligations. The most important elements of the plan were approved at the 6th session of the Gore-Chernomyrdin commission in Moscow in late January 1996. In order to reduce the number of Progress resupply missions to Mir, NASA agreed to fly an additional two Shuttle missions to Mir in 1998 to deliver about 6,000 kg of supplies and scientific experiments and to study the possibility of adding additional flights in 1999. In addition to that, the bulk of the Russian Science Power Platform components were now to be put into orbit in late 1999 on a single Shuttle mission instead of three Zenit launches and several of the less critical Russian add-on modules would be allowed to slip in the assembly schedule. In return, Russia pledged to deliver the FGB and the Service Module on time to ensure that the initial stages of station construction remained on schedule. As later events would show, the Russians were not able to keep that promise.

The Backbone: Zarya and Zvezda

FGB/Zarya

Design Features

Being the first element of the ISS, the FGB was to provide electrical power, attitude control and computer commands to the embryonic station and also to keep it in the proper orbit. It was to play an active role in docking the Service Module to the station and after this operation would mainly serve as a fuel depot and storage facility.

FGB is the Russian acronym for Functional Cargo Block (*Funktsionalno-Gruzovoy Blok)*, although in NASA documentation it was referred to as the 'FGB Energy Block'. In June 1998 it was officially named Zarya ('Dawn'), symbolising the beginning of a new era in space exploration, although this was hardly a new name in the Russian space programme [34]. A proposal by the Russian Space Agency to name it Atlant had earlier been rejected by NASA.

Zarya is a descendant of the Transport Supply Ships (TKS – index 11F72) originally designed in the late 1960s at the Chelomey bureau to ferry crews and supplies to the military Almaz space stations. Built at Khrunichev, the TKS vehicles consisted of a Functional Cargo Block (index 11F77) designed by the Fili branch of the Chelomey bureau and a Return Apparatus (index 11F74) designed by the central office of the Chelomey bureau in Reutov. The FGB section had storage room, propellant tanks and engines and the Return Apparatus was to house cosmonauts during launch, docking and landing. Four TKS vehicles were eventually flown, one on a solo mission (Cosmos-929), one to Salyut-6 (Cosmos-1267) and two to Salyut-7 (Cosmos-1443 and 1686). A modified FGB section called PGO (for Instrument Cargo Compartment) later served as the front section of the Mir modules Kvant-2, Kristall, Spektr and Priroda (common index 77KS). Attached to the PGO were one or more additional compartments tailored to each module's specific needs.

The lineage of Zarya is reflected in its index 77KM (serial number 17501). The vehicle is divided into two main sections, the Instrument Cargo Compartment (PGO) and a spherical docking hub referred to by the Russians as the Hermetic Adapter (GA). The PGO is subdivided into three compartments: PGO-2 is the 'aft' conical compartment with an active SSVP-M docking port for the Service Module and PGO-1 and PGO-3 are the two cylindrical compartments that constitute the pressurised core of Zarya. PGO-3 is separated by a hatch from the GA, which has an axial APAS-89 docking port compatible with Node-1/Unity and an SSVP nadir docking port. The zenith port was dropped early in the module's design. Outwardly, Zarya most closely resembles Kristall, the major difference being that on Zarya the pressurised core has the same diameter throughout **(Fig. 16)**.

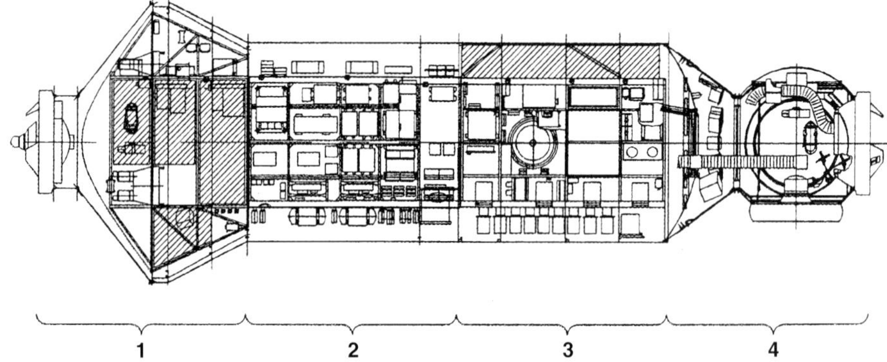

Fig. 16 The main compartments of Zarya.
(source: Novosti Kosmonavtiki)

Key: 1. PGO-2;
2. PGO-1;
3. PGO-3;
4. GA.

Zarya's engine system consists of two 417 kg thrust rendezvous and orbital manoeuvering engines, twenty-four 40 kg thrust approach and stabilisation engines and 16 Vernier engines with 1.36 kg of thrust each for attitude control. Exclusively used before the arrival of the Service Module, these engines drew their propellant from 16 tanks mounted in pairs on the exterior of the pressurised core (eight nitrogen tetroxide tanks and eight UDMH tanks). Ten of the tanks (five oxidizer, five fuel) hold propellant at high pressure for the small-thrust engines and the remaining tanks hold propellant at low pressure for the two high-thrust engines. After the arrival of the Service Module the propellant is stored for use by the Service Module's engines.

Electricity is provided by two solar arrays with a surface of 28 m² each (7 m in length and 4 m in diameter). Extending from the PGO-3 compartment, they have a combined capacity of 6 kWt. If needed, they can be folded up later in the station assembly process if they become shaded by other station elements. There are six nickel-cadmium batteries inside the PGO-1 compartment which store electricity for use during the nighttime portions of the station's orbit. Installed behind panels inside the PGO-1 and PGO-3 compartments are other FGB support systems as well as lockers for equipment storage, leaving only a relatively narrow passageway for crew members floating through the module. Zarya's command and control post is situated in the conical PGO-2 section and contains two US-built multiplex-demultiplex processors using Russian software [35].

The Road to the Launch Pad

NASA considered for a while to lease the FGB from the Russians, but given the critical nature of the vehicle in station assembly eventually decided to procure the vehicle from Khrunichev. Under the $400 million contract signed between NASA and the Russian Space Agency on 23 June 1994, $25 million **was earmarked for early work on the FGB. The contract to buy the FGB from Khrunichev was to be negotiated by Lockheed Missiles and Space Co., which had strong ties with Khrunichev through a joint venture called International Launch Services, responsible for marketing the Proton rocket. The money for the FGB was to flow from NASA to US station prime contractor Boeing to Lockheed to Khrunichev. It took several months to reach agreement on the total value of the contract and on the issue of who would pay for the launch. NASA found Khrunichev's selling price of $245 million unacceptable, with the Russians countering that the same piece of hardware built in the US would cost $1 billion. A final deal was not reached until early February 1995. NASA would pay an additional $190 million (beyond the $25 million approved in June) for the manufacture and testing of the FGB, while the Russian side would pay for the launch, in-orbit maintenance and flight control. Later that year Boeing became concerned about cost growth and management efficiency with Lockheed sandwiched between it and Khrunichev and therefore inherited the contract from Lockheed, with the signing ceremony taking place on 15 August 1995. Later NASA would pay an additional $35 million for modifications to the FGB, bringing the total value of the contract to about $250 million.**

Construction of the FGB flight model got underway in December 1994. With launch set for November 1997, Khrunichev had only about 2.5 years to complete assembly and testing before shipment of the FGB by rail to Baykonur in May 1997. This would only be possible thanks to the vehicle's strong lineage with the TKS spacecraft and Mir modules. Perhaps one of the greatest challenges from Khrunichev's perspective was to take into account the 500 or so specific design requirements levied upon it by the US, some of which were related to Zarya's 15-year design lifetime (much higher than that of the Mir modules). For instance, NASA insisted on six times more meteoroid and space debris shielding on the vehicle's hull than the Russians had initially proposed.

Assembly of the FGB went very smoothly, with the hull being completed by the end of 1995. There was a setback in early December 1995, when the internal bulkhead and part of the pressurised section were damaged during a pressure test, but this caused no significant delays. In keeping with tradition, Khrunichev built various mock-ups and test articles of the FGB and even a self-financed back-up vehicle (FGB-2 or 77KM N°17502) that could be sent into orbit within a year should something go wrong with the launch of the primary FGB.

By the spring of 1997 the FGB was essentially ready to be shipped to Baykonur, but problems with the Russian-financed Service Module had delayed the start of station assembly from November 1997 to June 1998. The Service Module was not only to act as the living quarters for the crew, but was also supposed to take over station reboosting from the FGB once it arrived in December 1998. Any further delays in the SM launch might cause problems, because the FGB had limited fuel supplies on board and could only be refuelled by Progress vehicles *via* the Service Module. Therefore an alternative had to be found for station attitude control and reboosting in case the SM slipped further once the FGB was already in orbit. NASA proposed an Interim Control Module (ICM) based on a classified Naval Research Laboratory upper stage dispenser used to manoeuvre and deploy Navy ocean surveillance satellites and even drew up plans for a follow-on Propulsion Module should the SM be delayed even further. Khrunichev on the other hand suggested to launch the FGB-2 as a temporary replacement for the SM, but RKK Energiya and the Russian Space Agency felt that this would have a negative impact on the SM launch preparations. One reason why the FGB-2 idea was not found attractive may have been that the vehicle would have to be modified with American money and therefore be considered by many as yet another US element of the station. Eventually it was decided to solve the problem by modifying the FGB such that it could control the station for an extended period of time, while also keeping open the option of launching the ICM should the need arise. A protocol on the FGB modifications was signed between NASA and the Russian Space Agency on 22 February 1997.

The major change required to the FGB was to allow it to be directly refuelled by Progress vehicles, which necessitated some changes to its nadir docking port. This port was a 'hybrid' SSVP-M type, capable of receiving the Docking and Stowage Module, which in turn could receive a Soyuz-TM or US crew rescue vehicle. The nadir port was now changed into a standard probe-drogue SSVP type compatible with Progress and equipped with the necessary connections for refuelling operations. Further, the FGB would be capable of storing 6.1 tonnes rather than 5.7 tonnes of propellant, although it would be launched with no more than 3.8 tonnes of propellant. Changes were also made to the FGB's control system so that it could handle more ISS configurations and to its axial SSVP-M port to make it compatible with the ICM.

The modifications to the FGB were completed by the end of 1997 and in January 1998 the module was shipped to Baykonur, where it was placed in the former Buran assembly building. In May 1998 further problems with the Service Module forced the launch to be delayed from 30 June to 20 November 1998. After mating with its Proton rocket, Zarya was rolled out the launch pad on 16 November and launch took place as scheduled on 20 November at 06.40 UTC, marking the long-awaited beginning of ISS construction (Fig. 17). About a month before the launch the Russians had unexpectedly proposed to delay the launch until 15.20 UTC so that Zarya would wind up in the same plane as the Mir station, making it possible for Progress vehicles to transfer some 2.1 tonnes of scientific equipment from Mir to ISS. This proposal was made despite the fact that long before this US and Russian orbital mechanics experts had settled on a launch time that would place the FGB into an orbit as far *away* from Mir as possible so that passes over Russian ground stations would not overlap [36]. NASA rejected the new launch time, saying it would require too many changes to the timeline for the upcoming STS-88 Shuttle mission that would dock Node-1/Unity to Zarya. It may also have seen the Russian move as a veiled attempt to revive the 1995 joint Mir/ISS assembly plan should the Service Module fall victim to further delays.

Fig. 17 Zarya in orbit as seen by the STS-88 crew.   (source: NASA)

Service Module/Zvezda

Design Features

The second module of the Russian segment serves as the cornerstone for early habitation of the ISS and also provides important flight control and orbit maintenance functions. Originally known as the Mir-2 Base Block, it was renamed the Service Module, which was not the best of choices. The same term had been coined in the 1960s for the propulsion compartment of the Apollo spacecraft and was accordingly often used in Western literature for the analogous section of the Soyuz spacecraft. Moreover, it was identical to the name of one of Mir-2's add-on modules, which had also been part of the ISS in the original Russian proposal. In May 1999 the Service Module was officially called Zvezda ('Star'), a name which had been used for at least five cancelled Soviet space projects in the past [37].

Having been originally built as a back-up for Mir, the Service Module is very similar in layout to the Mir core module. It consists of four major compartments: the Transfer Compartment (PKhO), the Work Compartment (RO), the Transfer Chamber (PrK) and the Instrument Compartment (AO) **(Fig. 18)**.

The spherical Transfer Compartment has three passive SSVP-M docking ports (axial, zenith and nadir) and can also serve as an airlock for EVAs until the arrival of the Docking Compartment. The Work Compartment is the main living and working zone for the crew and has two sections, one with a diameter of 2.9 m and the other with a diameter of 4.1 m. The first section among other things houses the central command post of the Service Module and provides room for installing the TORU remote-control docking system. The second section has two personal sleeping quarters, a personal hygiene facility, cooking facilities and a refrigerator-freezer, a table for securing meals while eating, equipment for physical exercises and a small airlock for disposing of garbage or ejecting small satellites.

The Transfer Chamber is a small tunnel allowing cosmonauts to transfer from the Work Compartment to an attached Soyuz or Progress spacecraft. It is equipped with a standard passive SSVP docking port, identical to the ones used on Mir. Surrounding the Transfer Chamber is the unpressurised Instrument Compartment, which contains the engines and propellant tanks of the SM's Combined Engine Installation (ODU). The ODU comprises two main correction engines (each with a thrust of 315 kgs) and a total of 32 attitude control engines (each with a thrust of 13.3 kgs). The propellant tanks can be topped up by Progress ships docked to the Zvezda aft port or to the Zarya and Zvezda nadir ports (or via any modules docked to the nadir ports). Propellant can also flow from the SM's tanks to the FGB tanks (for storage), back to Progress for additional manoeuvres and to the attitude control engines of the Universal Docking Module. The ODU main engines can only be used when the SM's aft docking port is unoccupied, meaning that most ISS orbit corrections are carried out by Progress vehicles docked to the aft port. However, the SM thrusters can be fired for attitude control during such Progress burns. Attached to the Instrument Compartment is a high-gain antenna, which makes it possible to communicate with the ground via Altair geostationary relay satellites.

Fixed to the outside of the 2.9 m diameter section of the Work Compartment are two solar panels, each of them covered with 38 m² of solar cells. Their total capacity is 9.8 kWt and this can be increased to 13.8 kWt with the attachment of four small add-on panels. Inside the SM are eight storage batteries, five of which are launched with the SM and three delivered later. The SM is supposed to provide power to the entire Russian segment (except the FGB) until the arrival of the Science Power Platform. The SM's power system can also load the storage batteries of Soyuz and Progress vehicles attached to the ISS. **In order to augment meteoroid and space debris protection cosmonauts will make several spacewalks to install 23 protective covers on Zvezda's hull as well as four shields on the solar arrays.**

The Service Module's brain is the DMS-R (Data Management System of the Russian Service Module), a set of on-board computers, their avionics and software that provide overall control, mission and failure management for not just the SM, but for the entire Russian segment. It is also responsible for the exchange of data and commands with other parts of the ISS. The key hardware for the DMS-R was developed in Europe by an industrial consortium led by

Fig. 18   Cut-away drawing of Zvezda (docking adapter and engine compartment not shown).
(source: Novosti Kosmonavtiki)

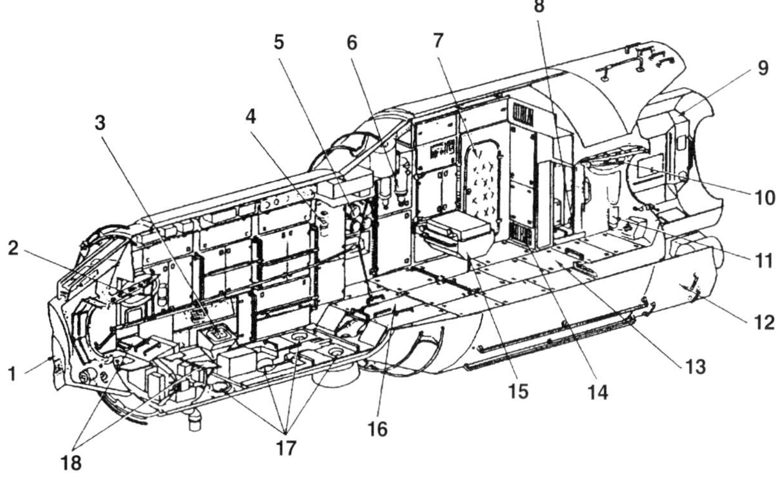

Key:
1. Transfer Compartment;
2. Hatch;
3. TORU remote-control docking equipment;
4. Gas mask;
5. Air cleansing units;
6. Solid-fuel oxygen generators;
7. Sleeping quarters;
8. Toilet;
9. Transfer chamber;
10. Hatch;
11. Fire extinguisher;
12. Instrument Compartment;
13. Treadmill attachment structure;
14. Dust collector;
15. Table;
16. Stationary bicycle attachment structure;
17. Portholes;
18. Central command post.

Astrium in Bremen, Germany. Actually, ESA had originally proposed the Data Management System as one of its contributions to the Mir-2 project in 1992. The ESA-provided on-board units are two Fault Tolerant Computers (a Control Computer and a Terminal Computer) and two control posts for command and control by the crew and for commanding experiments and the European Robotic Arm, which is to be installed on the Science Power Platform. The application software for the on-board computers was developed by RKK Energiya [38].

The Road to the Launch Pad

Although construction of the Service Module as part of the Mir-2 project had begun in the early 1980s, it would be responsible for the biggest delays in the start of ISS construction. This was mainly due to the fact that, unlike the FGB, it was to be funded entirely by the cash-strapped Russian government. The funding problems became evident by early 1996 and prompted repeated appeals from US officials to the Russians to honour their space station commitments or face the possibility of being elbowed out of the project. Despite Russian pledges to resume long-delayed payments to Khrunichev, RKK Energiya and their subcontractors, the Service Module was about 8 months behind schedule by the end of 1996. What had been obvious for many months became official in May 1997, when the Space Station Control Board decided to delay the start of station assembly from November 1997 to June 1998, with the Service Module slipping from April to December 1998. Exactly one year later further delays in funding and building the Service Module forced yet another postponement of its launch to April 1999, with the FGB moving to November 1998.

An important milestone was reached on 1 June 1998, when the Service Module was finally shipped from Khrunichev to RKK Energiya's ZEM factory for final outfitting and electrical tests. An option that had been considered to save time was to transport it directly from Khrunichev to Baykonur and finish all the necessary work there (as had been done with the Mir core module in 1985), but this was considered too risky. By early October ongoing funding problems, compounded by the financial crisis that hit Russia in August of that year, necessitated a postponement of the SM launch to July 1999, although the start of station construction in November 1998 remained on schedule. Work on the SM gained fresh impetus that same month with a $60 million cash infusion from NASA, in return for which the Russian Space Agency agreed to give NASA up to 75 % of its research crew time and extra stowage space aboard the SM during the assembly phase. This was supposed to be only the first payment under a $660 million bailout plan crafted by NASA to help the Russians complete crucial hardware for the ISS until 2002. It was to be followed by four annual payments of $150 million, mainly to help defray the cost of building Progress and Soyuz vehicles for resupply and reboost during assembly. However, NASA later decided against including the remaining $600 million in its budget request because it would take the pressure off the Russian government to contribute its share of money to ISS.

In May 1999 the Service Module was shipped to the Baykonur cosmodrome, where the final hardware was to be installed. Another slip in the SM launch date was announced in June, this time from July to 12 November 1999. In late September the space station partners agreed on a delay to sometime between 26 December and 16 January, mainly because the Space Shuttle fleet was temporarily grounded for wiring inspections after a potentially dangerous short-circuit during the launch of STS-93 in July. Zvezda's launch was to be followed by a Shuttle outfitting and resupply mission and managers wanted the interval between the two flights to be as short as possible. Other reasons for the delay were software integration problems between computers on the US segment of the station and the European Data Management System on the Service Module as well as delays in preparing Russian ground stations for the launch.

A major setback came on 27 October 1999, when Zvezda's launch vehicle, the Proton rocket, suffered a second-stage engine failure during the launch of an Express communications satellite from the Baykonur cosmodrome. It was the second Proton failure in just under four months, with an almost identical accident having taken place on 5 July during the launch of a Gorizont communications satellite. Although the Proton had made two successful flights in September, this latest failure under very similar circumstances required a thorough analysis, throwing the Proton launch schedule into complete disarray. In early January 2000 the investigation board concluded that just like the July accident the 27 October failure had been caused by contamination in one of the 2nd stage engines. Not coincidentially, the engines from both ill-fated flights were from a batch manufactured at the Voronezh Mechanical Plant back in 1992-93, when the facility had been in the midst of a production slump. The commission recommended to modify the 2nd and (nearly identical) 3rd stage engines by installing filters and using special nickel coating in the turbompumps to protect them from burn-throughs. On 11 February the Council of Chief Designers met to discuss the impact on the Zvezda launch and decided to reschedule it for 8-14 July on condition that it was preceded by two successful Proton launches with modified 2nd and 3rd stage engines. The latest delay was officially included in the space station assembly schedule in late March.

Two partially modified Protons (using only the filters) were successfully launched on 12 February and 17 April, but the first Proton with all the recommended modifications did not fly until 6 June. After two more launches of partially modified Protons on 24 and 30 June, Zvezda was transported to a Baykonur facility where its tanks were to be loaded with propellant. This critical and irreversible operation was not to begin until after the launch of the second completely modified Proton early on 5 July, which came on the brink of failure after experiencing a drop of pressure in its 2nd stage fuel tank. Considering this a random problem, the Russians gave the green light for fuelling Zvezda *before* informing the other space station partners about the mishap. With no way of turning back, Zvezda was mated with its Proton launch vehicle on 6 July and rolled out to the pad two days later. What was probably the most critical launch in the history of the ISS took place on 12 July at 04.56.36 UTC (Fig. 19). For Zvezda it was the end of an almost 25-year countdown that had begun with the approval of the DOS-7 and DOS-8 space stations way back in 1976. The Russians took every necessary precaution not only to make sure that the launch went off without a hitch, but also to ensure that Zvezda safely docked to Zarya/Unity. Just in case something went wrong, a two-man rescue crew (Padalka/ Budarin) was on stand-by to fly to Zvezda aboard a Soyuz spacecraft and dock Zarya to Zvezda using the TORU remote-control manual docking system. In the event the module successfully docked with Zarya/Unity on 26 July at 00.44.44 UTC, making it possible for ISS assembly to begin in earnest.

Fig. 19 Launch of Zvezda on 12 July 2000.
(source: NASA)

The Add-On Elements

Since the whole ISS assembly sequence hinged on the availability of the FGB and Service Module, virtually all attention in the 1990s was focused on putting these cornerstones into orbit on schedule. With the Russians barely finding the necessary funds to build Zvezda, the development of the add-on elements was put on the backburner for many years. Since these were pretty much out of the 'critical path', there was much less pressure on the Russians to launch them on time, which probably explains why they were allowed to go through a myriad of design changes as the years went on. Even after the launch of Zarya and Zvezda, the future of the add-on elements remains uncertain and only one of them (the Pirs Docking Compartment) has been launched to date.

Original Plans

According to the ISS assembly plan agreed upon in September 1994 all the Russian elements to be attached to the FGB and the Service Module were to be both designed and built by RKK Energiya and delivered to the station by a propulsion compartment that would be detached after docking. The Docking Compartment would be orbited by the Soyuz rocket and ferried to the station by a standard Progress-M propulsion compartment, while all the other elements would go up on Zenit and use the Progress-M2 propulsion compartment as their space tug. After the jettisoning of their propulsion compartments, most of the modules would have an aft docking port available for receiving other vehicles.

The design of all these modules can easily be traced back to Mir-2. The Docking Compartment (which doubled as an airlock for EVAs) remained largely unchanged from its Mir-2 configuration, although it would now receive Soyuz and Progress ships rather than Buran. The Universal Docking Module, made up of three compartments and sporting six docking ports, was also virtually identical to its Mir-2 progenitor, although another option studied was to make it smaller and orbit it with a Soyuz rocket [39]. The Docking and Stowage Module, the Life Support Module and the three Research Modules outwardly looked pretty much the same as the other Mir-2 add-on modules, consisting of a single 6.5 m long compartment weighing about 8 tonnes.

Although the design of the DSM and LSM was inherited from the Mir-2 modules, both were introduced specifically for ISS. The DSM, essentially a storage facility for equipment ferried up by Progress resupply ships, was probably conceived after the negative experience with cluttering aboard Mir, where cosmonauts would often spend long hours searching for lost items. In addition to this, the DSM had an aft docking port for Soyuz-TM ships and later American crew rescue vehicles. The LSM was to have several life support systems similar to those used on Mir, such as an Elektron unit for oxygen production and a system to recycle water from urine. It was also to carry various personal hygiene facilities including a small sauna. Preliminary plans for the Research Modules called for a technological, a biotechnological and a remote sensing module. One option studied was to use the Euro-Russian Technological Complex (see Mir-2 section) as one of the Research Modules.

The Science Power Platform was a simplified version of the cross-beam originally intended for Mir-2, which in its 1992/1993 configuration had two solar parabolic concentrators and a set of four solar panels. In the original ISS plans the Russians had intended to install the concentrators on the SPP, but by late 1993 they were forced to use the four solar panels instead. In 1994 the number of panels was increased to six. According to the September 1994 assembly plan the SPP was to be built up in various stages using specialised Progress-M2 vehicles in which the payload would replace the 6.5 m long pressurised cargo compartment (the 'assembly version' of Progress-M2 described in the Mir-2 section). The first such mission was to carry up the SPP-1 section, consisting of the lower truss, a pressurised section for gyrodins and storage batteries and a radiator panel. After delivering these elements to the zenith port of the Service Module, the Progress-M2 propulsion compartment would separate, opening a docking port for the SPP-2 section, which was to be transported to the station in similar fashion. SPP-2 comprised the middle and upper truss, a solar array drive installation, a Portable Engine Unit (VDU – similar to the one carried on Mir's Sofora truss) and the European Robotic Arm (ERA) plus a rail structure allowing the arm to move along the SPP. Several spacewalks would be required to deploy the various elements and to extend the ERA rail all the way down to the Universal Docking Module. After that two or three more specialised Progress-M2 vehicles would dock with the UDM to deliver the solar panels and their attachment structures plus additional or replacement VDU thruster packages. These were to be transferred to the SPP using the ERA or a cargo crane mounted on the Service Module. The gyrodins for the SPP-1 section were to be sent up separately aboard several standard Progress-M or Progress-M2 cargo ships [40] (Fig. 20, 21).

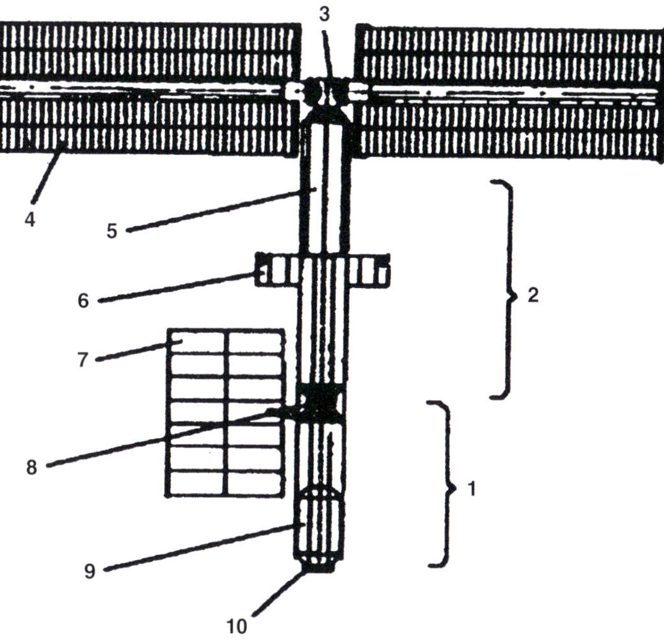

Fig. 20 An early version of the Science Power Platform with four solar panels. (source: Novosti Kosmonavtiki)

Key:
1. SPP-1 section;
2. SPP-2 section;
3. Solar array drive mechanism;
4. Solar panel;
5. Extendable part of SPP-2 section;
6. VDU thruster package;
7. Radiator;
8. SPP-1/SPP-2 interface;
9. Pressurised section with gyrodins;
10. SPP-1/Zvezda interface.

Fig. 21 Cosmonauts using the ERA to install a solar array. (source: Fokker Space)

Zenit Dropped

The configuration of the add-on elements remained unchanged until the 6th meeting of the Gore/Chernomyrdin commission in late January 1996, where it was decided that the Zenit rocket would not be used in station assembly until at least 2000. Several months later the Russians dropped the Zenit from the ISS assembly schedule altogether. The official reasons given by Yuriy Koptev after the Gore/Chernomyrdin meeting were the high risks associated with having only one Zenit launch pad at Baykonur (the other one had been destroyed in a pad explosion in October

1990) and the relatively high cost of the rocket. Another reason must have been that Russia wanted to avoid cash payments to Ukraine, where the Zenit was assembled [41].

All this automatically meant that the Progress-M2 based ISS modules scheduled to fly to ISS would have to be cancelled. The implications were reflected in Revision B of the ISS assembly schedule, released in September 1996. The Universal Docking Module would now be built on the basis of FGB-2, the Zarya back-up. The DSM and the LSM were each split into two modules (DSM-1, 2 and LSM 1, 2) to be docked in tandem to the FGB nadir port and one of the UDM radial ports respectively. Outwardly similar to the Mir Docking Module, each module would weigh about 3 tonnes after jettisoning its propulsion unit and launch would take place with either the Soyuz or the upgraded Soyuz-2. Only two Research Modules remained by this time, and consideration was given to building them on the basis of the FGB design. The SPP was to be delivered by two Shuttle missions (9A.1 and 14A) and would now have a total of eight solar panels with a capacity of 50 kWt. The first Shuttle mission was to carry the truss components, the pressurised section with gyrodins, four solar panels and the ERA arm and the second another set of four solar panels. In 1998 the delivery of the final four solar panels was split between two Shuttle missions (1J/A and 14A) to make room in the cargo bay for some of Zvezda's anti-debris shields. Transferring the SPP components from the cargo bay to the Zvezda zenith port required a complex operation in which the Shuttle's RMS robot arm would hand over the elements to the station's SSRMS manipulator arm.

Since it would take a certain amount of time to transform the FGB-2 into the UDM, its launch slipped considerably. This left the Russian segment without adequate EVA capability in the early stages of assembly, because the airlock-equipped Docking Compartment was to use the UDM as its berthing place. This is why it was decided to build two Docking Compartments. The first one would link up with Zvezda's nadir port and be deorbited by an attached Progress vehicle when the UDM was ready to take its place. Subsequently, a new DC would be launched to dock with the UDM. A more economical approach would have been to redock the first DC after the arrival of the UDM (using the same Progress vehicle), but this was not possible because of the use of several types of docking ports on the Russian segment. While the Zvezda nadir port had an SSVP-M port, the UDM's aft docking adapter only had SSVP ports, meaning the original DC could not be redocked to the UDM [42].

More Changes

In 1997 the Khrunichev Centre and RKK Energiya began design work on the FGB-2 based UDM. Although most of the work on the module would be done at Khrunichev, RKK Energiya remained the prime contractor. As in the earlier Progress-M2 design, the primary function of the module was to receive add-on modules at its aft axial port and then transfer them to one of the four radial ports. The aft port would also be able to receive Soyuz and Progress vehicles, with the latter being able to refuel the Zarya and Zvezda propellant tanks via the UDM. Aside from housing scientific equipment and a computer for controlling the Russian segment, the UDM would also carry the gyrodins originally intended to be installed on the SPP and use its engines to provide roll control. The propellant for that would be fed from the Zvezda fuel tanks (Fig. 22).

Instead of constructing the UDM on the basis of FGB-2, it was soon decided to build it from scratch, because FGB-2 would have required too many modifications. For instance, the UDM's docking adapter needed five docking ports, while FGB-2 had only two, and it also had to be capable of acting as back-up airlock in case cosmonauts wouldn't be able to seal the DC-2 hatch after an EVA. Unlike FGB-2, the UDM was not supposed to perform the role of fuel depot and power facility, requiring only four fuel tanks (compared to sixteen on FGB-2)

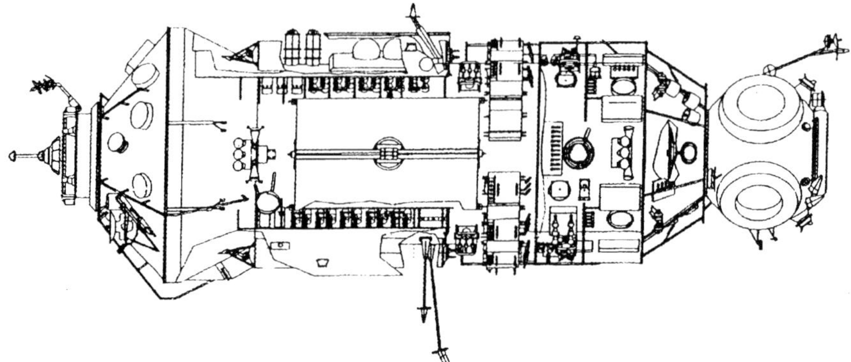

Fig. 22 The Universal Docking Module based on the FGB design.
(source: Novosti Kosmonavtiki)

and small non-rotatable solar panels that would be folded after the module's arrival at the ISS (meaning that the bays for the solar array drive mechanisms in FGB-2's hull had to be removed). The installation of the gyrodins would also have required major changes to the FGB-2's interior.

The first half of 1998 saw several more significant changes to the UDM design. The multiple docking adapter was to be turned 45° clockwise and the nickel-cadmium storage batteries were to be replaced by nickel-hydrogen batteries that would be mounted outside rather than inside the module, freeing up 2.5 m³ of internal volume. Other changes resulted from a decision made by the Council of Chief Designers on 28 April 1998 to scrap both Life Support Modules and return to a single large Docking and Stowage Module which would now be built on the basis of FGB-2. The UDM's six gyrodins were transferred to the Docking and Stowage Module and the additional room that became available inside the UDM could now be used to install some of the equipment originally intended for the LSMs (including a system to recycle water from urine, the Elektron oxygen generator, a small washing facility and a mini-sauna). These latest changes to the Russian segment were reflected in Revision D of the ISS assembly schedule, released in May 1998. Design work on the UDM was temporarily halted in July-August 1998 because of a lack of financing and also because RKK Energiya and Khrunichev could not agree on the exact division of work. A final agreement between the two was not signed until October 1998.

The new FGB-based DSM retained its original function of equipment storage facility, but was now also to play an important role in the station's attitude control. Not only did it now carry the six gyrodins, its engines were also to take part in attitude control, drawing propellant from Zvezda and Zarya. In addition to that, special propellant lines were to be routed through the DSM to allow Progress vehicles docked to the DSM's aft port to refuel Zarya. Two radial docking ports would be available on the aft docking adapter, although the purpose of those was not announced.

Meanwhile, plans for the two Research Modules remained as uncertain as ever. An important development came on 31 May 1997, when Russian and Ukrainian presidents Boris Yeltsin and Leonid Kuchma signed an agreement on space co-operation, one aspect of which was the inclusion of a Ukrainian research module in the Russian segment. This meant that there would now be one Russian and one Ukrainian Research Module. However, with Ukraine having no experience in building space station modules, the contract for the Ukrainian module would have to go to either RKK Energiya or Khrunichev.

At the meeting of the Council of Chief Designers in April 1998 RKK Energiya and Khrunichev presented competing proposals for the Russian Research Module, although both probably were eying the Ukrainian contract as well. Despite the 1996 decision to refrain from the use of the Zenit in ISS assembly, RKK Energiya once again came up with a Zenit-launched Progress-M2 look-alike vehicle weighing 9 tonnes (minus the tug) which could later be outfitted with 3 additional tonnes of scientific equipment. Khrunichev presented a DOS-based vehicle weighing 20 tonnes at launch and 24 tonnes after being outfitted in orbit. However, it reportedly also proposed a stripped-down FGB-based vehicle that was compatible with Zenit, making it more attractive to the Ukrainian side. In both proposals the modules had standard scientific racks, making it easier for the international partners to gain access to them. The Ukrainian space agency was expected to pick one of the proposals in early 1999, but no such decision followed, probably because Ukraine could not afford the $100 to 150 million price tag associated with building the module.

Later TsNIIMash, Russia's main civilian space R&D insitute, released its long-awaited specifications for the Russian module's scientific equipment, which would require a module weighing 16 tonnes. Responding to this announcement, Khrunichev donwsized its module to a 16-tonne vehicle that would be towed to the station by a 4 tonne, 2.9 metre diameter detachable space tug using FGB hardware. This was somewhat reminiscent of the configuration of Kvant, although the mass distribution between module and tug on that vehicle had been significantly different (11 tonnes for the module and 9 tonnes for the tug). The module proper would consist of a conical section derived from the FGB design and a 4.1 m diameter cylindrical section inherited from the DOS space stations. Other configurations were proposed as well, including smaller modules with externally mounted scientific platforms. Since RKK Energiya had an entire department responsible for scientific equipment aboard space stations, Khrunichev proposed that it would solely be responsible for building the module and tug and then hand over the module to Energiya after it reached ISS.

However, Energiya continued to pursue the idea of building its own modules and in May 1999 presented a plan to the Russian Space Agency for a unified series of Zenit-launched vehicles with identical propulsion compartments. This basically marked a return to the Progress-M2 concept for Mir-2 and the initial version of the ISS Russian segment. Although the May 1999 plan centred on a NASA-funded replacement for the Docking and Stowage Module, Energiya suggested to use the same design for the Russian and Ukrainian Research Modules and later also for a heavy cargo ship. An alternative considered for the Zenit rocket was the all-Russian Yamal, an uprated Soyuz rocket for which RKK Energiya was the lead design bureau. Using improved engines in the first and third stages and an NK-33 engine in its core stage, Yamal had a payload capacity only slightly less than that of the Zenit. Reviewing the proposals for the Research Modules on 1 July 1999, the Russian Space Agency decided that both the RKK Energiya concept (using either Zenit or Yamal) and Khrunichev's FGB/DOS-based module/tug combination (using the Proton-M) required further study. A follow-up meeting to decide on a final design never took place, leaving the future of the modules up in the air [43] (Fig. 23).

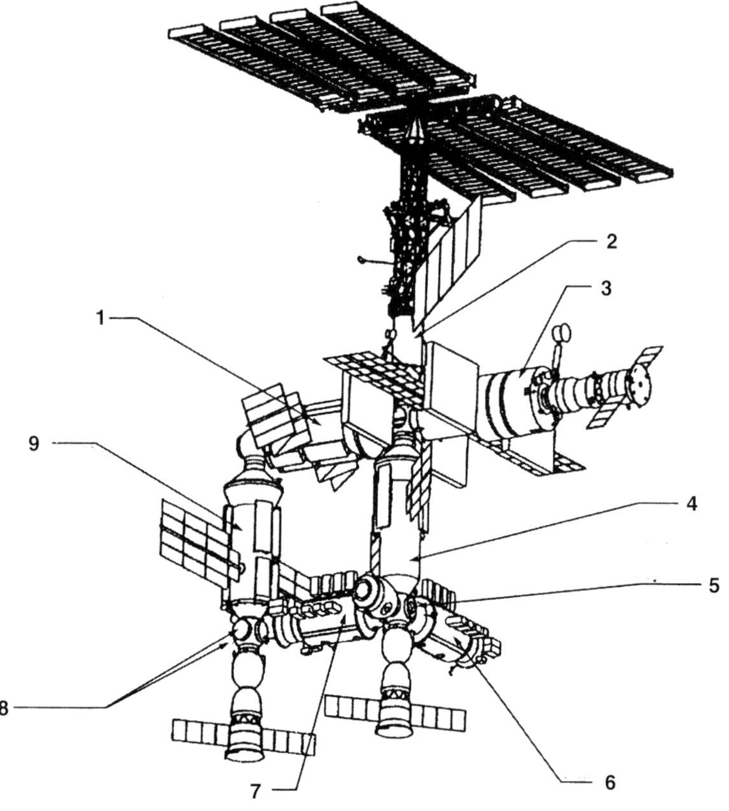

Fig. 23 The Russian segment as planned in late 1998.
(source: Novosti Kosmonavtiki)

Key: 1. FGB;
2. Science Power Platform;
3. Service Module with anti-debris shields on the solar panels;
4. Universal Docking Module;
5. Free docking port;
6. Research Module 1;
7. Research Module 2;
8. Two free docking ports;
9. Docking and Stowage Module.

Going Commercial: Enterprise and the CSM

As 1998 drew to a close, the only add-on module that looked like it had a chance of flying anytime soon was the nearly completed FGB-2, but the prospects of turning it into the Docking and Stowage Module looked dim with the scarce government funds available. Khrunichev had already spent a large amount of its own resources on the vehicle and clearly wanted to gain some commercial benefits from it. There were ideas to use it as a Logistics Transfer Vehicle and even as a free-flying remote sensing module in the framework of a proposed European Global Environmental Service. Part of the problem with deciding the fate of FGB-2 was that no commitments could be made until after Zarya was safely launched into orbit. On 21 November 1998, the day after the Zarya launch, Khrunichev and Boeing signed a memorandum on the possibility of using the FGB-2 in a different role than Zarya back-up.

Khrunichev's first proposal was to turn FGB-2 into what it called a Multipurpose Module (Russian abbreviation MTsM). Intended as a replacement for the Docking and Stowage Module, it could house American scientific equipment originally supposed to be installed in the US Laboratory Module Destiny and the US/Japanese Centrifuge Accommodation Module. Because of delays in the launches of these modules, NASA had already bought space and crew time aboard Zvezda to run some of the US experiments earlier, but the proposed Khrunichev module would offer much more room to carry out the experiments. Negotiations on adapting the FGB-2 for that role began in February-March 1999 and in May 1999 Khrunichev and Boeing officials held two weeks of talks in the US which were expected to end with the signing of a contract under which NASA would buy the module in an arrangement similar to the purchase of Zarya (with Boeing acting as an intermediary). However, just two days before the end of the talks, Boeing received a letter from RKK

Energiya with a suggestion to build the MTsM on the basis of its newly proposed series of unified vehicles and launch it with a two-stage version of the Sea Launch Zenit (Zenit-2SL) from Baykonur. Boeing officials had already been informed about this option by RKK Energiya at a meeting of the Sea Launch partners on the Cayman Islands on 29-30 April. Since Boeing was the leading partner in the Sea Launch venture, it found RKK Energiya's offer attractive and asked for two months to study the proposals of both Khrunichev and Energiya. In a letter to Yuriy Koptev on 13 May 1999 Yuriy Semyonov outlined some of the advantages of RKK Energiya's vehicle over the FGB-2, claiming it could better perform the tasks originally set aside for the Docking and Stowage Module and underlining that the same design could be used for the Russian and Ukrainian Research Modules. A Russian Space Agency meeting the following day concluded that both proposals deserved further study, but in early July Boeing informed its Russian partners that NASA had abandoned the idea of building the MTsM for financial reasons [44].

With little hope in sight of securing any significant Russian government funding for the expansion of the Russian segment, both RKK Energiya and Khrunichev continued to pursue the idea of selling their module concepts to Western customers. After negotiations held in July and October 1999, RKK Energiya signed a contract with Washington-based Spacehab Inc. on 8 December 1999 to jointly develop a commercial ISS module. In the contract the vehicle was referred to as the Commercial Docking and Stowage Module (Russian abbreviation KMSS), but two days later it was announced to the press as Enterprise. The Russians also continued to call it the Multipurpose Module, since it was essentially the same vehicle offered to NASA/Boeing earlier in the year, although now it would have accommodations for microgravity experiments like the ones Spacehab had been flying for paying customers on the Space Shuttle for many years. Enterprise would also be equipped with a tiny broadcast studio to produce educational and entertainment programming that would be distributed on a commercial basis via television and the Internet. The costs for Enterprise's development were expected to be split in half between Energiya and Spacehab, with each partner providing $50 million, although other Western companies such as DASA and Mitsubishi were expected to join later. On 11 April 2000 Spacehab announced that it had formed a new subsidiary called Space Media Inc. that would have exclusive rights to selling the information beamed down from Enterprise. The original idea was to launch Enterprise with the Yamal rocket, but in April 2000 RKK Energiya decided to substitute it for the Zenit, hoping that this would lead to some kind of deal with the Ukraine on the construction of that country's Research Module. Since the Zenit had a longer payload fairing than Yamal (17.65 m as compared to 14.80 m), it also became possible to lengthen both the module and the tug. As a result the tug could carry more propellant and the pressurised volume of the module grew from about 35 m³ to 45 m³, 25 m³ of which was available for installing cargo [45].

Meanwhile, Khrunichev was still setting its sights on Boeing to join in the development of its own commercial ISS module, now aiming for a direct contract with Boeing without NASA involvement. The two companies reached an agreement in early May 2000 and an official announcement was made at the Farnborough air show on 27 July. Called the Commercial Space Module (CSM), the vehicle would be built on the basis of FGB-2. It would first deliver up to 3 tonnes of fuel and cargo to ISS as a prototype of the Logistics Transfer Vehicles and then offer about 20 m³ of internal volume for installing scientific and communications equipment [46].

Although it was clear that both Enterprise and the CSM were supposed to become part of the Russian segment, there was confusion as to where exactly the modules would dock. First, the Zarya and Zvezda nadir ports were officially still occupied by Russian government modules (the DSM and UDM resp.) and second, the information provided by Energiya/Spacehab and Khrunichev/Boeing indicated that *both* wanted to have their modules attached to the Zarya nadir port. Things became somewhat more transparent with a joint press release from Energiya and Spacehab on 8 August 2000, stating that Energiya president Yuriy Semyonov and Russian Space Agency director Yuriy Koptev had signed a preliminary agreement on 17 May 2000 on the replacement of the DSM by Enterprise. Although NASA had been informed about this decision at the time, the news was not made public until several days after the CSM announcement and was undoubtedly timed to serve as a reminder to Boeing/Khrunichev that the Zarya nadir port was not up for grabs. If it were to come down to a race between the two modules, the CSM was certainly at an advantage, because FGB-2 was already 70 % ready and was expected to be launched by mid-2002. Enterprise, on the other hand, existed only on paper and was not scheduled for launch until early 2003. The 17 May agreement also stipulated that the Russian Space Agency would pay for the Zenit launch and that RKK Energiya and Spacehab would compensate for this by using some of the revenues earned with Enterprise to 'improve the characteristics of the Russian segment, such as energy, communications and other resources'. This would have included the installation of additional solar panels on the Service Module and the deployment of a geostationary communications satellite called Passat to relay data to and from Enterprise and the remainder of the Russian segment [47].

In September 2000 the design of the Zenit-launched Enterprise was finalised, but in early October it became known that the price charged by the Ukrainian Yuzhnoye design bureau for the Zenit launch would be the same as that in the Sea Launch programme, forcing RKK Energiya to look at other launch options. By the end of October the choice fell on the more powerful Proton rocket and two months later the design of Enterprise was adapted accordingly. Although the overall dimensions of Enterprise and its tug changed little, the module would now be able to carry up to 3.2 tonnes of internally mounted equipment into orbit as compared to 500 kg when launched by Zenit. This would significantly reduce the number of supply missions to Enterprise, which was supposed to have 4.85 tonnes of internally mounted equipment when fully outfitted. The module (minus tug) was now 2.9 m in diameter, 8.87 m long and provided 48 m³ of pressurised volume. As before, it internally consisted of two sections, one for storage and scientific experiments and the other an in-orbit television studio for live broadcasts. There was also room for up to 1.5 tonnes of externally mounted experiments. An aft docking port was available for Soyuz and Progress vehicles and provisions were made for installing gyrodins and refuelling Zarya and Zvezda via Enterprise. In this way, Enterprise retained the major functions of the cancelled Docking and Stowage Module, as had been the intention from the outset [48]. A major milestone was reached on 5 March 2001, when Energiya and Spacehab announced that they had reached a final agreement with the Russian Space Agency on 16 February to include Enterprise in the Russian segment and dock it to the Zarya nadir port. In late February the Russian Space Agency had sent an official request to NASA to adapt the ISS assembly schedule accordingly [49].

With the Zarya nadir port now firmly booked, the Enterprise deal seemed to sound the death knell for the CSM. However, on 13 April 2001 Boeing and the Russian Space Agency signed a wide-ranging agreement on co-operation in space and aviation, one aspect of which was to study the commercial use of the FGB-2 module. Although this seemed to contradict the Enterprise deal, Yuriy Koptev said that a way would be found of including both modules in the Russian segment. For the moment though, the future of the CSM hung in the balance [50].

Pirs Joins the Russian Segment

Meanwhile, preparations were in full swing for the launch of the Docking Compartment, the first add-on element of the Russian segment. The DC has its roots in a docking adapter built in the late 1980s for Buran (Fig. 24). Similar in concept to the later Orbiter Docking System of the US Space Shuttle, the docking adapter was to allow Buran to perform dockings with other spacecraft and act as an airlock for EVAs during docking missions. It consisted of a spherical section (2.55 m in diameter) topped by a cylindrical tunnel (2.200 m in diameter) with an APAS-89 docking port. The spherical section, bolted to the floor of the cargo bay, had two side hatches, one to connect it with Buran's middeck and the other to provide access to a Spacelab-type module or allow cosmonauts to perform EVAs. The tunnel provided the actual interface between the docking adapter and the target vehicle and would be extended to its full length after the opening of the payload bay doors. With the tunnel fully extended, the adapter was 5.7 m high. At least one flightworthy version was built for a planned docking flight of the Buran-2 orbiter with a Soyuz-TM spacecraft and the Mir space station.

An adapted version of Buran's docking adapter was included in the Mir-2 and later in the ISS design to receive spacecraft (Soyuz, Progress and originally also Buran) and to serve as an airlock. It would be towed to the station by a detachable Progress-M propulsion compartment. The central part of the spherical section (2.55 m in diameter) was retained. Mounted to its aft end was a small section of the Buran airlock's cylindrical tunnel (without the extendable part) and attached to the front end was the forward part of a Soyuz-TM/Progress-M orbital module (Fig. 25). The design was more complex than that of the Docking Module of Mir (316 GK), which did not have to be used as an airlock and was merely an extension to the Kristall module to facilitate Shuttle dockings [51].

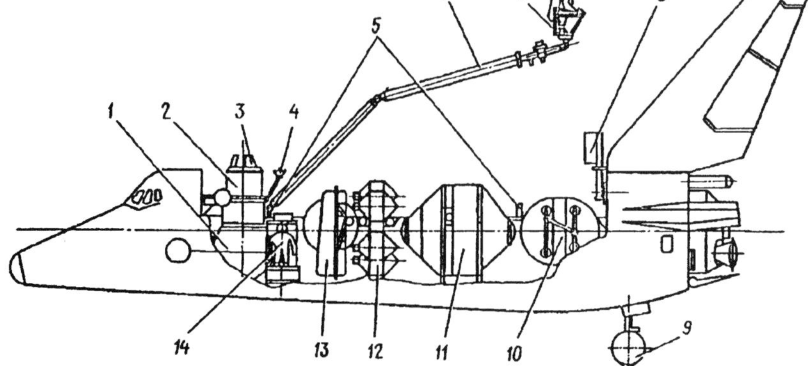

Fig. 24 : The Buran docking adapter (position 3), the progenitor of Pirs.

(source : RKK Energiya)

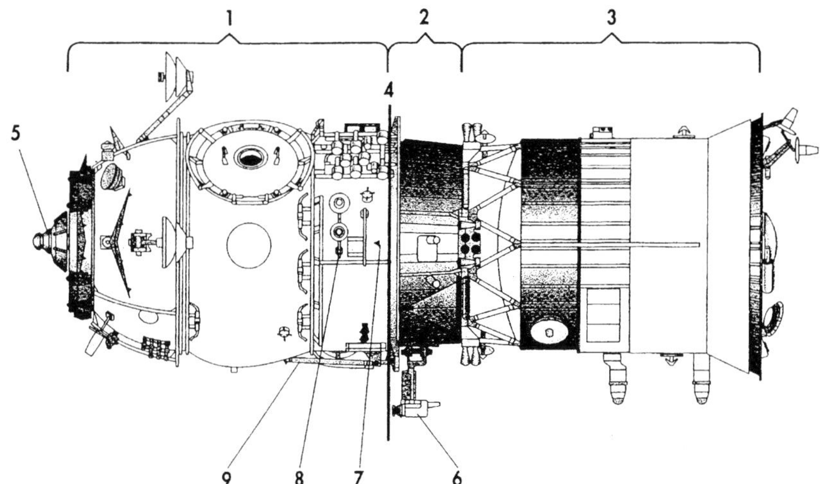

Fig. 25 : Drawing of Progress M-SO1.
(source: Novosti Kosmonavtiki)

Key: 1. Pirs Docking Compartment;
2. Intermediate section;
3. Propulsion compartment;
4. Interface between Pirs and intermediate section;
5. SSVP-M docking port;
6. Television camera;
7-9. Antennas.

In the final ISS design the front docking port is an active SSVP-M (compatible with Zvezda's nadir port) and the aft port a passive SSVP (compatible with Soyuz and Progress). Progress vehicles can refuel Zvezda via the DC. There is a 1-metre diameter EVA exit hatch on either side of the spherical section, each having a small porthole in the centre. The two exit hatches date back to the Mir-2 design, where they would have enabled cosmonauts to assemble parts of the Science Power Platform inside the module with pieces sticking out from both sides. The choice of the hatches is now determined by which of the two provides the easiest access to the work area for a particular EVA. Both hatches are inward opening. Although this reduces the working volume inside the airlock, it precludes an eventuality where the hatch is violently swung open if there is any residual pressure inside the airlock. Such a situation caused permanent damage to the outward opening hatch of Mir's Kvant-2 module during a spacewalk in July 1990.

The DC is connected to its propulsion compartment via a short intermediate section, which is pyrotechnically separated from the module along with the propulsion compartment after arrival at the ISS. The propulsion compartment itself is almost identical to that of the Progress-M spacecraft, containing an engine unit, propellant tanks, solar panels and housekeeping systems needed during the vehicle's autonomous flight.

Construction of the module began at RKK Energiya in 1998 and was finished in late 2000. Joint electrical tests with the Progress-M based propulsion compartment got underway in late January 2001 and were finished in late June 2001 (Fig. 26). On 16 July the vehicle was delivered to the Baykonur cosmodrome, where it underwent final preparations in the former Buran assembly building. By this time the Docking Compartment had been dubbed 'Pirs' ('pier'). Even this was not a new name in the Russian space programme, having been used earlier for a cancelled ocean reconnaissance satellite system [52]. Among the designers the compartment was known as 240GK or SO-1 (for *Stykovochnyy Otsek* or Docking Compartment). The combination of the Docking Compartment and its Progress-M propulsion compartment was called Progress M-SO1, which is how the vehicle was officially registered after its launch with a Soyuz-U rocket on 14 September 2001 at 23.34.55 UTC (15 September local time) [53]. After a standard two-day Progress rendezvous profile, the spacecraft docked with Zvezda's nadir port on 17 September at 01.05.14 UTC. The propulsion compartment was jettisoned on 26 September at 15.36 UTC and was de-orbited on 27 September at 23.30 UTC.

Among the approximately 800 kg of cargo installed in Pirs was an extra Orlan-M spacesuit and a 'Strela' cargo crane (GStM-2) to facilitate cosmonauts' movements on the exterior of the complex. Another 'Strela' (GStM-1) had already been carried up on two Shuttle missions and temporarily mounted on one of the American Pressurised Mating Adapters by spacewalking astronauts. Both Strela booms were installed on opposite sides of Pirs during spacewalks in October 2001 and January 2002 [54].

The Currently Planned Configuration

By the time Pirs was launched, there was some more clarity about the future of the Russian segment, notably about the commercial modules. Negotiations on another reconfiguration of the Russian segment had been held in June-July 2001 and resulted in the signing of a joint document on 8 August by Koptev, Semyonov, TsNIIMash director N. Anfimov and Khrunichev's new director A. Medvedev. The proposed changes were approved by the Council of

Fig. 26 Progress M-SO1 ready for shipment to Baykonur.   (source: RKK Energiya)

Chief Designers on 28 August. After Pirs the following elements were to be added to the Russian segment: the Commercial Space Module on the basis of FGB-2, the Enterprise Multipurpose Module, a simplified Science Power Platform and two Research Modules (Fig. 27). The plans were slightly adjusted the following months and this is the planned assembly sequence for the remainder of the Russian segment as of mid-2002.

The next element to join the Russian segment should be the Boeing/Khrunichev Commercial Space Module, expected to be launched in early 2004. It will occupy the Zvezda nadir port instead of the Universal Docking Module. Actually, it will take over many of the functions of the UDM and is therefore also referred to by the Russians as the Simplified UDM. Under a contract signed between Khrunichev and Boeing on 27 December 2001, Boeing will pay the $50 million needed to adapt the FGB-2 for its new role. Khrunichev had earlier spent $53 million from its own resources to build FGB-2 as a back-up for Zarya. The European Astrium consortium may also join the deal. Some matters still have to be resolved with the Russian Space Agency before the work can go ahead, such as the thorny issue if the agency will finance the Proton launch. The CSM will have four docking ports, a front axial SSVP-M port to dock with Zvezda, an aft axial SSVP port to receive Soyuz and Progress vehicles and two radial SSVP ports on the aft docking adapter to receive the two Research Modules. Eight of FGB-2's sixteen fuel tanks will be removed from the outer hull, making room for some equipment that was originally supposed to be installed on the SPP. This includes batteries for converting and distributing the electricity produced by the SPP's solar panels and a part of the SPP's radiator panel to dissipate the heat produced by the batteries. Boeing and Khrunichev expect to gain revenue by providing room for experiments to paying customers. For this purpose a set of scientific racks will be installed inside the module with standard interfaces also used on the Destiny module.

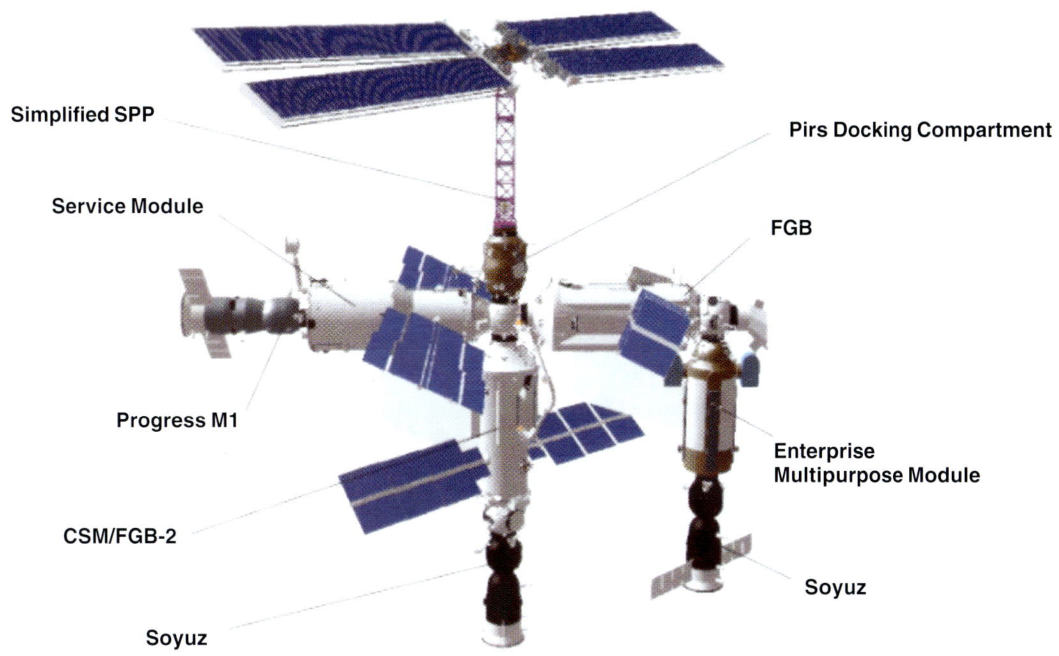

Fig. 27 The currently planned configuration of the Russian segment.   (source: Spacehab)

TABLE 1: *Evolution of launch dates for elements of the ISS Russian segment. The November 1993 schedule is the baseline schedule, followed by several revisions made over the years. The final column lists the expected launch dates at the time of writing, which are subject to change.*

	Nov 1993	Sep 1994 (Rev. A)	Sep 1996 (Rev. B)	Sep 1997 (Rev. C)	May 1998 (Rev. D)	Jun 1999 (Rev. E)	Aug 2000 (Rev. F)	Jul 2002 (unofficial)
FGB	5/97	11/97	11/97	6/98	11/98	launched 11/98	launched 11/98	launched 11/98
SM	7/97	4/98	4/98	12/98	4/99	11/99	launched 7/00	launched 7/00
DC (1)	6/97	7/98	5/99	12/99	3/00	9/00	3/01	launched 9/01
DC 2	-	-	5/00	12/00	5/01	7/02	8/03	-
UDM	8/97	6/98	4/00	12/00	4/01	6/02	8/03	-
SPP-1*	9/97 (A)	11/98 (R)	11/99 (A)	7/00 (A)	1/01 (A)	11/01 (A)	10/02 (A)	7/04 (A)
SPP-2*	10/97 (R)	2/99 (R)	6/01 (A)	5/02 (A)	10/01 (A)	10/02 (A)	2/04 (A)	10/04 (A)
SPP-3*	-	5/99 (R)	-	-	8/02 (A)	8/03 (A)	5/05 (A)	1/06 (A)
DSM (1)	-	1/00	3/01	2/02	3/02	7/03	no date	-
DSM 2	-	-	no date	5/02	-	-	-	-
LSM (1)	6/99	10/00	2/02	1/03	-	-	-	-
LSM 2	-	-	3/02	3/03	-	-	-	-
RM 1	2/01	9/99	5/01	8/02	8/02	3/04	8/05	8/05
RM 2	8/01	6/00	12/01	11/02	11/02	8/04	3/06	4/06
RM 3	9/01	5/01	-	-	-	-	-	-
CSM	-	-	-	-	-	-	-	?/04
Enterprise	-	-	-	-	-	-	-	7/04 (A)

Notes: *Elements of the Science Power Platform have been scheduled for launch by both Russian (R) and American (A) launch vehicles. SPP-1, SPP-2 and SPP-3 refer to different components of the SPP at different times. Note that the September 1994 plan actually required a total of five Russian launches.

Shortly before the arrival of the CSM, the station's SSRMS robot arm will be used to transfer the Pirs Docking Compartment from the Zvezda nadir port to the Zvezda zenith port, which will become its permanent location on the ISS. Prior to this transfer an EVA will have to be conducted to install an SSRMS grapple fixture on the exterior of

Pirs. Moving Pirs to the Zvezda zenith port will no longer make it necessary to discard Pirs and launch a new Docking Compartment, as envisaged earlier. Pirs' aft docking port will now serve as the berthing place of the SPP, which has been significantly simplified. It will consist of a single extendable frame with four instead of eight solar panels (capacity 25 kWt instead of 50 kWt) and the European Robotic Arm with its rail structure. The radiator panel, the pressurised section for the storage batteries and the thrusters have been removed.

The resulting reduction in size and mass of the SPP left a lot of extra room on the Shuttle mission that was supposed to deliver most of its components to the ISS. Therefore Spacehab and RKK Energiya came up with a plan in October 2001 to launch the SPP *together* with the Enterprise module on a single Shuttle mission. Until then Enterprise had been slated for launch on a Proton, but since the module is a competitor for the CSM, Khrunichev was not expected to sell a Proton to Energiya/Spacehab at the going Russian domestic rates. If launched on the Shuttle, Enterprise would also no longer require a space tug to tow it to the ISS. Still awaiting approval, the current plan is to launch the entire SPP (the truss, four solar panels and the ERA) together with Enterprise on Shuttle assembly mission 9A.1 in July/August 2004. Four additional SPP solar panels currently remain manifested for missions 1J/A and 2J/A in 2004 and 2006 (two panels per mission) and may therefore be installed after all. Using both the Shuttle's own manipulator arm and the station's SSRMS arm, the SPP elements will be mounted on top of the Pirs module and Enterprise will be attached to the Zarya nadir port.

Enterprise may become the key to expanding the station's resident crew from three to six. Faced with massive budget overruns on ISS, NASA was forced in 2001 to indefinitely delay the Habitation Module and the Crew Rescue Vehicle (CRV), which effectively reduced the station's permanent crew to three and thereby severely limited the station's science capabilities. At the end of the year an independent task force headed by former Martin Marietta president Thomas Young recommended three ways of giving the station the needed life support capacity to house six crew members on a permanent basis. One was to build a habitat module on the basis of the Multipurpose Pressurised Logistics Modules (MPLM), a second to equip Node-3 with US life support systems and a third to outfit Enterprise with Russian life support systems. The first option foresaw the use of the CRV with a considerable financial contribution from ESA, while the second and third options involved the purchase of Russian Soyuz rescue craft. RKK Energiya and Spacehab had already begun exploring the possibility of turning Enterprise into a habitat module in February 2001 and the negotiations to launch the module on the Shuttle were apparently part of this effort. The exact financial details of any possible deal are still unclear. RKK Energiya and Spacehab have indicated that they intend to rent Enterprise to the space station partners "in a package" with the Soyuz spacecraft, which would serve as a lifeboat for the additional crew members working on board Enterprise.

Some significant changes have been made to the module to adapt it for its possible new role. In the latest design Enterprise is 2.9 m in diameter and 9.2 m long with a pressurised volume of 50 m³. Although the ability to produce television programmes will remain, the dedicated broadcast studio has been replaced by three individual crew cabins and a toilet. The module will also house some of the life support systems originally intended for the Russian Life Support Module, such as an Elektron oxygen generator, a Vozdukh carbon dioxide scrubber and a system to recycle water from urine. There will still be room for scientific equipment (in both Shuttle type lockers and Destiny type racks), but far less than in the original plans. Installed on the exterior will be standard interfaces for Spacehab's Integrated Cargo Carrier (ICC) platforms, another part of the SPP's radiator panel as well as several body-mounted radiators and a set of gyrodins to take part in station attitude control. The gyrodins, originally supposed to be mounted on the SPP, then on the UDM and later on the DSM, will be delivered by Progress cargo ships and installed during spacewalks. Enterprise will retain its two SSVP docking ports, a front port for linking up with Zarya's nadir port and an aft port for receiving Soyuz and Progress vehicles (Fig. 28).

The assembly of the Russian segment should be rounded out with the addition of the two Research Modules in August 2005 and April 2006. Plans for a Ukrainian-funded module have apparently been given up, although the Russian and Ukrainian space agencies did agree in February 2002 on a series of 48 Ukrainian experiments to be conducted on the Russian segment in the coming years. No decision has been made yet on the configuration of the two modules and if no funding turns up they may never be launched at all [55].

Free-Flyers

Russia has also announced plans for free-flying spacecraft that would orbit in the vicinity of the ISS and periodically dock with it. The rationale behind this is that many experiments are much easier to perform from a free-flying

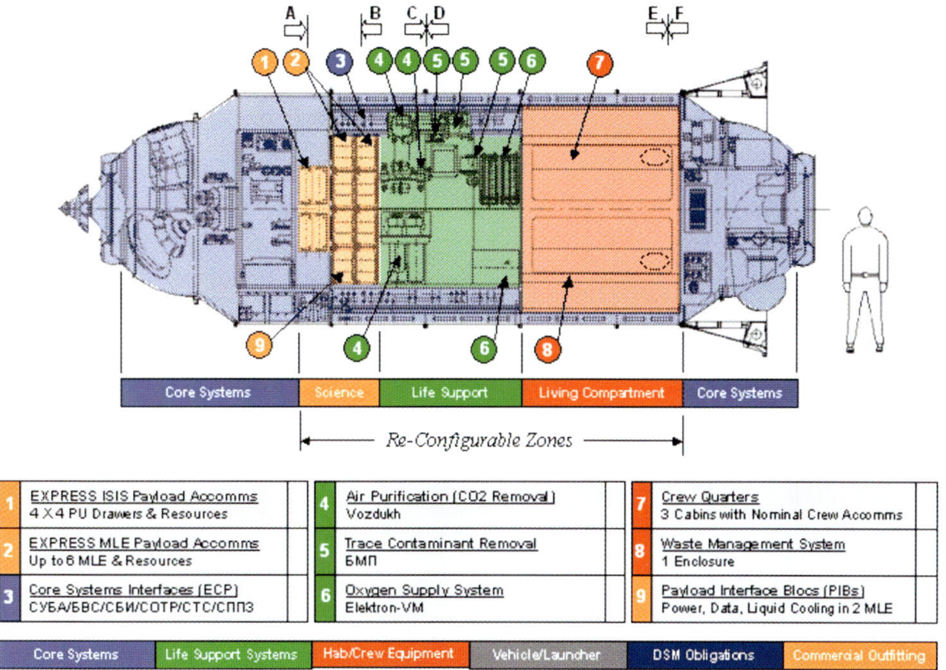

Fig. 28 The currently planned configuration of Enterprise   **(source : Spacehab)**

platform than from a multimodular space station. This is especially the case for sensitive materials processing experiments, which require a vibration-free environment, and astronomical observations, which demand accurate pointing of telescopes. The idea is that such spacecraft regularly dock with the station for maintenance, refuelling, upgrading and both loading and unloading of scientific equipment. Plans for such free-flyers have been around for a long time. For instance, back in the early 1970s there were plans in the Soviet Union for a giant N-1 launched space station called MKBS which would have been joined by several such co-orbiting free flyers. An outgrowth of these was the Progress-based Gamma astrophysics module, launched in 1990.

The first proposal for a free-flyer was put forward by TsNIIMash as early as 1996 and is called MAKOS-T (for Reusable Automatic Space Orbital System - Technological) (Fig. 29). This is a spacecraft for materials processing experiments which exclusively uses proven technology. It consists of three sections: an engine unit and equipment bay (with housekeeping systems) both derived from the Phobos/Mars-96 space probes (developed by NPO Lavochkin) and a cargo module identical to that of the Progress resupply ships (developed by RKK Energiya). MAKOS-T is designed to remain in orbit without refuelling for three years and is supposed to dock with the ISS Russian segment twice a year, making it possible to run six cycles of technological experiments. Adapted versions of this spacecraft could also be used for astronomical observations, remote sensing of the Earth and particles and fields studies. In 1999 the Russian Space Agency was seeking partners interested in financing MAKOS, but the current status of the project is not clear.

Between 1997 and 2000 RKK Energiya worked on a free-flyer called AKA-T (for Automatic Space Apparatus – Technological). Weighing 7.8 tonnes, it had the same overall dimensions as Soyuz/Progress and aside from a pressurised cargo compartment had an airlock chamber to expose materials to the vacuum of space. It was able to carry up to 800 kg of materials processing equipment and was supposed to visit ISS two to three times a year. One of the payloads would have been a multifunctional furnace called Karat-T to manufacture various types of crystals using a wide range of production methods. One version of AKA-T had a boom with a shield extending from its aft compartment. The shield, which created a near-perfect vacuum in its wake, was to employ a technique called molecular beam epitaxy to produce ultrathin layers of semi-conductor materials for the microelectronics industry. Developed as part of a programme known as Ekran, this experiment was similar to those performed with the US Wake Shield Facility, a free-flyer flown on three Space Shuttle missions between 1994 and 1996.

A more recent proposal for a free-flyer was made jointly by RKK Energiya and the Astro-Space Centre of the Physical Institute of the Academy of Sciences and calls for the development of an astrophysics platform called SLK (Free-Flying Spacecraft) (Fig. 30). The SLK has a Progress-type pressurised module and aft propulsion compartment, but the usual propellant compartment is replaced by an unpressurised section with an infrared telescope called Submillimetron. A similar Progress-based infrared observatory (Aelita) was proposed by NPO

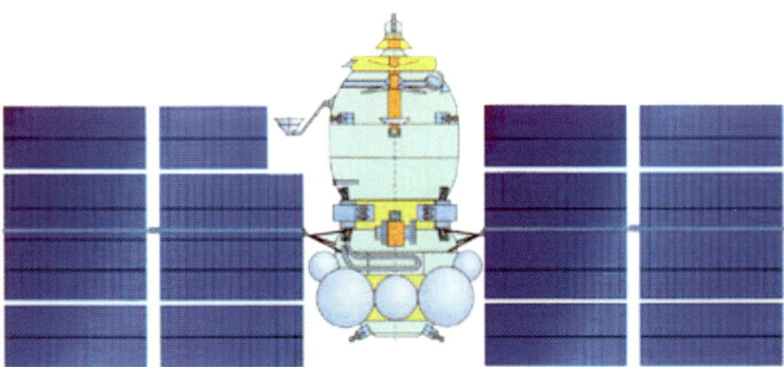

Fig. 29 MAKOS-T. (source: Novosti Kosmonavtiki)

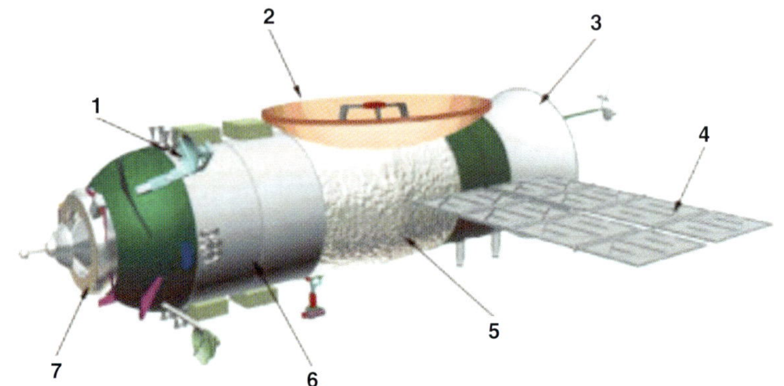

Fig. 30 The SLK. (source: Novosti Kosmonavtiki)

Key: 1. Docking system antennas;
2. Telescope mirror;
3. Propulsion compartment;
4. Solar panel;
5. Unpressurised equipment compartment;
6. Pressurised compartment;
7. Docking port.

Energiya in the 1980s. The Submillimetron telescope could be used to make a complete infrared survey of the sky and would focus among other things on galaxies and quasars, galactic sources of star formation, interstellar and interplanetary dust. It could also study minor fluctuations in the cosmic background radiation and spot potentially dangerous near-Earth asteroids. The SLK could remain in orbit for as long as ten years and would periodically dock with ISS, among other things to replenish the liquid helium used to cool the telescope. However, as in the case of MAKOS, private funding will have to be found to turn this idea into reality [56].

The Ferries

Soyuz

Soyuz as Part of Freedom

In the wake of the Challenger accident NASA issued a request of proposals in 1987 for what it called an Assured Crew Return Vehicle (ACRV) for space station Freedom. Since the Space Shuttle could only temporarily remain docked to the station, another means had to be found to bring astronauts back to Earth in case of an emergency inbetween Shuttle visits. The tasks formulated for the ACRV were to return an ill or injured astronaut or to bring back an entire crew if the station were to become uninhabitable or if the Space Shuttle was grounded indefinitely. Freedom could not be permanently manned until such a lifeboat became available. Both Lockheed and Rockwell studied a number of options (including lifting bodies and Gemini and Apollo-based vehicles), but all of those were expected to cost between $1 and 2 billion and would take several years to develop.

With the collapse of the USSR imminent and a new international political climate taking shape, NPO Energiya managers saw an opportunity to come to NASA's rescue. In October 1991 Energiya's general designer Yuriy Semyonov met with Boeing vice president R. Grant at the Congress of the International Astronautical Federation in Montreal, Canada to discuss among other things the possibility of using a Soyuz type vehicle as Freedom's ACRV. Semyonov repeated this offer while appearing at a Senate subcommittee hearing on 21 February 1992 and during the same week also continued his talks with Grant. This eventually led to a NASA delegation visiting Moscow in March to hold exploratory talks on the concept of using Soyuz as an ACRV. Besides Soyuz-TM, the Russians also proposed a vehicle derived from the cancelled 14F70/Zarya earlier envisaged for the giant Mir-2. This could house five to six astronauts and be orbited either by a Zenit rocket or in the cargo bay of the Shuttle.

On 18 June 1992, one day after the Bush/Yeltsin summit in Washington that paved the way for future US/Russian space co-operation, NASA and NPO Energiya signed a $1 million contract under which the two organisations among other things would jointly study the use of Soyuz-TM as an interim lifeboat for Freedom until the American ACRV became available to replace it. One of the problems to be tackled was how to put Soyuz into the 28° inclination orbit where Freedom was supposed to operate, something which was not achievable with the Soyuz launcher from Baykonur. The Russians proposed alternative launch schemes using Zenit or Proton launch vehicles, but these were rather complicated. The Proton would have to place Soyuz into a 51.6° parking orbit and then use a Blok-D upper stage to transfer it to the required inclination. Even then, the mass of Soyuz had to be significantly reduced, which is why the Russians asked NASA to increase Freedom's inclination to 33.5°. This in turn implied that the Shuttle's payload capacity to Freedom's orbit would decrease, which was unacceptable to NASA.

Other boosters considered were the Titan or Atlas from Cape Canaveral or even Ariane from Kourou. In the end though specialists settled for placing Soyuz in the cargo bay of the Space Shuttle. This required only a few modifications to the Soyuz, such as removing equipment needed for rendezvous operations and developing systems to firmly secure it in the Shuttle's payload bay. After the Shuttle's arrival at the station, the Soyuz could be transferred to the required docking port with the remote manipulator arm of the Shuttle or the station itself. The 28° inclination of Freedom also meant that Soyuz could not land in the traditional landing zone in Kazakhstan. Several alternatives were evaluated, the most promising of which was found to be Australia. In November 1992 a team of American and Russian specialists went to Australia and explored four different landing areas, three of which were considered to be acceptable.

Lockheed and Rockwell, the two companies vying to build the American lifeboat, were hoping to win the contract that NASA was expected to award for work on the interim Soyuz ACRV. In September 1992 both companies signed separate agreements with NPO Energiya to study Soyuz hardware. After two weeks of talks in Houston in December 1992 NASA and Energiya officials concluded that there were no major show-stoppers to adapting Soyuz-TM as a station lifeboat. In March 1993 NASA and NPO Energiya signed a new contract under which the Soyuz/Shuttle plan would be examined in more detail. The main focus would be on finding ways of extending Soyuz' on-orbit lifetime to 1-3 years, much more than the 6 months it was designed to stay in orbit. NASA officials were invited to witness the landing of the Soyuz TM-16 spacecraft in July 1993 to get acquainted with the recovery techniques. In December 1993 another round of talks was held in Houston to evaluate the results of the studies. Although the prospects of using Soyuz-TM as a Freedom lifeboat were good, NASA cancelled the contract with NPO Energiya that same month after the final decision to build the international space station jointly with the Russians. A competing proposal for Freedom's interim ACRV is said to have come from NPO Mashinostroyeniya, the former Chelomey design bureau. It involved the use of the three-man return capsule (VA) of the TKS vehicles. The bureau had reportedly even designed 6 and 8-man versions of the VA, but none of the proposals were found to be acceptable [57].

Soyuz as Part of the ISS

After the Russians joined Alpha in late 1993, there was no need to develop a specialised ACRV for the initial assembly phase. As long as the station was staffed by just three crew members, the Soyuz-TM could serve both as a crew delivery and rescue vehicle as it did in the Mir programme. With Alpha using the 'Russian' inclination of 51.6°, Soyuz-TM needed no modifications and could be orbited as usual by the Soyuz rocket. The station's resident crew was not scheduled to be expanded to six until after the arrival of the US Hab module in 2002. From that point on two Soyuz-TM vehicles needed to be permanently parked at the station, but given the restrictions of Soyuz and the high cost of regularly changing them out, NASA wanted to develop a larger and more capable vehicle that could evacuate the *entire* crew and remain docked at the station for several years.

Responding to these requirements, RKK Energiya and Rockwell (later joined by Khrunichev) drew up plans in 1995 for a rescue craft that looked like an enlarged Soyuz descent capsule with a small engine compartment at the back. It drew heavily on the 14F70/Zarya design of the late 1980s and on a concept that the Russians had already proposed for the Freedom ACRV in early 1992. The 8-tonne descent capsule had a maximum diameter of 3.7 metres and could house a maximum of eight crew members. With a total mass of 12 tonnes and a length of 7.2 metres, the vehicle could be delivered to ISS by the Shuttle and remain on stand-by for five years [58] (Fig. 31). However, the Russian proposal was not the only one. Europe also showed considerable interest in developing a crew return vehicle and in 1995 NASA started its own studies of a lifting-body type vehicle that could fulfil the role of ACRV. Eventually, NASA and ESA decided to join forces to build the lifting body, which became known as the X-38.

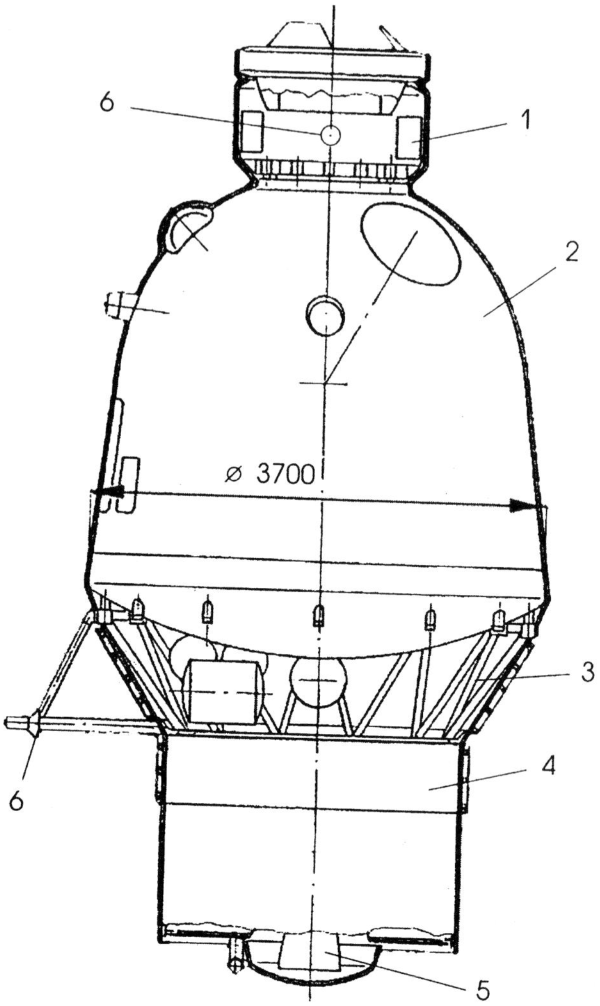

Fig. 31 "Big Soyuz" ACRV concept.   (source: RKK Energiya)

Key:
1. Front compartment with docking port;
2. Descent compartment;
3. Intermediate compartment;
4. Instrument and propulsion compartment;
5. Main engine (with protective cover);
6. Shuttle payload bay attachment structures.

At any rate, until the US lifeboat became available, Soyuz-TM would be the only means of escape for an ISS resident crew in case of an emergency. An issue that needed to be resolved were the very stringent requirements that the Russians had for a cosmonaut's body length to fit inside the cramped Soyuz descent capsule. The problem came to a head in late 1995 when two American astronauts, Scott Parazynski and Wendy Lawrence, had to be bumped from Mir training for being too tall and too short respectively. According to NASA figures published at the time 45 percent of the American astronaut corps could not fit within the Soyuz. Actually, NASA had been aware of the body length restrictions since the Soyuz lifeboat studies for Freedom. In order to slightly ease these restrictions Energiya engineers had proposed that the astronauts would not wear pressure suits in case of an emergency return aboard Soyuz. The risk was considered acceptable given the low probability of a crew evacuation and the fact that Soyuz would be delivered to Freedom by the Shuttle and only be manned during the return to Earth.

Around mid-1995 NASA asked RKK Energiya to re-evaluate the problem of using Soyuz as an emergency lifeboat for resident crews launched to ISS aboard the Shuttle. RKK Energiya again put forward a plan to return the crew without pressure suits, although a 'leak compensation system' would now be developed to allow the astronauts to survive a depressurisation of the descent capsule during an emergency return. Launch of the Soyuz to ISS would be unmanned. NASA, however, preferred to slightly modify the Soyuz so that it could accommodate more US astronauts *with* pressure suits, not only for safety, but also to avoid a situation where there would be dedicated 'rescue' and 'transport' versions of Soyuz. **Preliminary agreement on such modifications was reached at the 6th session of the Gore-Chernomyrdin commission in late January 1996.** The expectation was that the modifications would be made under a separate contract between NASA and RKK Energiya, but eventually it was decided to make an amendment to the existing June 1994 contract between NASA and the Russian Space Agency. By early July it was agreed that NASA would pay $39 million for the modifications and on 19 September 1996 the amendment to the contract was officially signed. A Draft Plan for the modified Soyuz was approved by Yuriy Semyonov on 24 December 1996. The vehicle was **called Soyuz-TMA, the A standing for 'anthropometric'.**

The major change in Soyuz-TMA is that the seat frameworks (in which the individually tailored seatliners are installed) have been enlarged. This will allow the vehicle to carry astronauts with a standing height of between 150-190 cm and a sitting height of between 80-99 cm (compared to 164-182 and 80-94 cm resp. for Soyuz-TM). In order to install the larger seats, small changes had to be made to the hull to accommodate the feet of the crew members sitting in the left and right seats. It was also necessary to redesign the control panel (which has new computer displays) and to relocate some equipment under the seats (Fig. 32). By installing new shock absorbers in the seats, the weight limits for the crew members have been expanded from 56-85 kg on Soyuz-TM to 50-95 kg on Soyuz-TMA. The difference in weight between the left and right crew member can be as large as 45 kg. The automatic landing system has been adapted to handle the resulting changes in the ship's centre of gravity. In another change, two of the six soft-landing engines have been redesigned so that they can be operated at two different thrust levels depending on the exact landing mass of the descent capsule (ranging from about 2,980 kg to 3,100 kg). Soyuz-TMA will have a maximum mass of 7,200 kg (about 200 kg more than Soyuz-TM) and will therefore be launched by a slightly uprated Soyuz rocket with improved first and second stage fuel injectors. Called Soyuz-FG, the new rocket made its debut with the launch of Progress M1-6 in May 2001 and is now considered man-rated [59]. Later on Soyuz-TMA is expected to be launched with a further uprated rocket called Soyuz-2. Soyuz-TMA went through an extensive test programme, which included four drop tests of mock descent capsules from an Ilyushin-76 plane at an Air Force base near Akhtyubinsk between December 1998 and December 1999. The first flight of Soyuz-TMA (serial number 211) is currently planned for the autumn of 2002 and it will mark the first time in Soyuz history that a new modification makes its maiden flight with a crew on board. Four more Soyuz-TMA vehicles (nrs. 212-215) are now in various stages of assembly at RKK Energiya for flights in 2003 and 2004 [60].

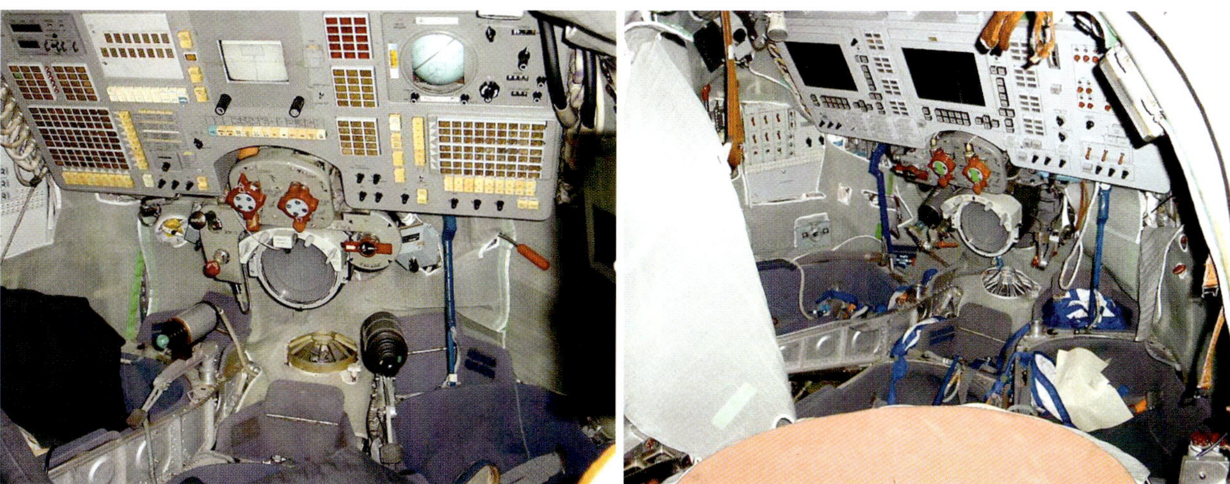

Fig. 32 Comparative views of Soyuz-TM (left) and Soyuz-TMA (right) cockpits.
(source: Mark Shuttleworth website at www.africaninspace.com)

RKK Energiya is already looking beyond Soyuz TMA and planning models that incorporate some more basic changes. RKK Energiya president Yuriy Semyonov had already issued an order to start work on improved Soyuz and Progress vehicles as early as November 1995. The goal would be to switch to modernised and lighter on-board systems manufactured solely in the Russian Federation (because companies situated in the former Soviet republics charged excessive prices), to enable various systems in the descent module to be reused and to increase the maximum flight duration to about one year. There was no room for such work under the Soyuz-TMA contract with NASA, which only covered the changes needed to accommodate taller astronauts. An offer by the Russians to include the work needed to increase the vehicle's lifetime had been turned down by the US space agency. It was not until June 1997 that Energiya signed a contract with the Russian Space Agency to start further modification work in earnest, with a Draft Plan for an improved model called Soyuz-TMM being approved in August 1998. The major modifications planned for Soyuz-TMM were:

- a single computer, installed in the descent module (as opposed to one in the descent module and one in the propulsion compartment on Soyuz-TM)
- an improved telemetry system
- two additional braking engines in the Approach and Docking engine system for safer dockings (probably a result of the collisions and near-collisions in the Mir programme)
- a satellite data relay system called Regul

- an autonomous satellite navigation system and an improved Kurs-MM rendezvous system

- deployment of the parachutes at a lower altitude and modifications to the automatic landing control system in order to increase landing accuracy

- increase of the maximum flight duration to 380 days. This will be achieved among other things by installing a thermoelectric cooling system for the tanks of the hydrogen peroxide thrusters on the descent module, improving the ship's batteries and making the oxidizer tanks in the propulsion compartment out of steel rather than an aluminium alloy.

It soon turned out that incorporating all these changes simultaneously would be too costly and by the summer of 1999 plans were approved for an intermediate version called Soyuz-TMS. This will only have the two extra braking engines, the improvements needed for more accurate landings (the eventual goal being to shift landings from Kazakh to Russian territory) and the cooling system for the hydrogen peroxide thrusters, although maximum flight duration will remain limited to 180-210 days. Soyuz-TMS will retain the separate computers in the descent module and propulsion compartment. The ship will also have an updated flight control system that among other things will increase docking safety and make the ship more attractive to NASA as a crew rescue vehicle. It will also have all the changes introduced by Soyuz-TMA. Semyonov approved the Draft Plan for Soyuz-TMS in December 1999 and if the necessary funds are made available, the vehicle could make its maiden flight in 2006 or 2007. The launch vehicle will be the Soyuz-2. Soyuz-TMM right now remains no more than a distant dream [61].

So far only the first ISS resident crew has used Soyuz to actually fly to the station. All subsequent resident crews went up and returned aboard the Shuttle. This means that the exchange of Soyuz vehicles has to be carried out by short-stay visiting crews on what have been called 'taxi missions'. These are now performed roughly every six months, whenever the maximum on-orbit lifetime of a Soyuz expires.

Due to serious cost overruns in the ISS programme it looks unlikely that the US-built Crew Rescue Vehicle (CRV) will ever fly. If the resident crew is increased to six, NASA will likely be forced to purchase extra Soyuz rescue craft, which was recommended in two of the three options for crew expansion presented by the Thomas Young commission in late 2001. In October 2001 the Russian Space Agency stepped up the pressure on NASA to opt for this solution by warning that foreign astronauts might soon lose long-term access to the ISS. The agency cited an 11 June 1996 agreement between the ISS partners under which Russia is obliged to deliver a Soyuz to the station only during the first 50 months of crewed operations. After that, the agency said, Soyuz would only be accessible to Russian cosmonauts unless the partners buy additional Soyuz lifeboats to provide emergency escape for non-Russian crew members [62]. Whatever the outcome of this debate, it looks like Soyuz will remain a key transportation system for the ISS for the foreseeable future.

Progress and the Logistics Transfer Vehicles

In the original station plans the ISS was to be resupplied and refuelled both by the Progress-M and Progress-M2 freighters. The Soyuz-launched Progress-M had been flying regular missions to the Mir station since 1989 and needed no modifications for its ISS role. The much more capable Zenit-launched Progress-M2 (see Mir-2 section) was originally conceived for Mir-2 and aside from serving as a cargo ship could also be adapted as a specialised tanker or as a tug for station modules or large structures to be used in station assembly. The cargo version was capable of delivering up to 5.7 tonnes of cargo and 800 kg of propellant to ISS and the tanker version up to 5 tonnes of propellant. Therefore, Progress-M2 was expected to allow the number of Progress-M missions to be significantly reduced in the course of station assembly.

All this changed with the decision made in early 1996 to drop the Zenit rocket (and consequently the Progress-M2 vehicles) from ISS assembly. After the 6th meeting of the Gore-Chernomyrdin commission in January 1996 it was announced that the indefinite delay of Progress-M2 was to be compensated by the development of several heavy cargo vehicles based on the FGB. Jointly financed by the US and Russia, the FGB-based freighters would be able to deliver about 10 tonnes of cargo, about as much as three Progress-M vehicles. NASA referred to these as Logistics Transfer Vehicles (LTV) and the Russians called them GTK-FGB (GTK standing for Cargo Transport Ship). Several of them were included in the ISS assembly schedules compiled later in 1996. However, in 1997 NASA decided not to fund the LTVs and instead carry up additional supplies on the Shuttle. Subsequently, Khrunichev proposed that the LTVs be funded by Russia alone to resupply only the Russian segment, but this drew heavy criticism from RKK Energiya, which did not want to see Russian money earmarked for its Progress craft to be transferred to the Khrunichev vehicles. Another argument used by RKK Energiya against the LTV were the relatively serious consequences of a launch or docking failure compared to the loss of a single Progress.

Little has been revealed about the exact modifications that would be required to the FGB to turn it into an LTV. One idea was to mount 22 propellant tanks on the outside of the pressurised hull and replace the aft docking adapter by a small undetachable space tug, similar to the one proposed at one point for the Research Modules. This space tug also featured in a proposed 'assembly version' of the LTV, where the pressurised module would be replaced by a cross-shaped platform outfitted with various types of equipment to be mounted on the exterior of ISS [63]. On several occasions Khrunichev proposed to turn FGB-2, the Zarya back-up, into an LTV demonstrator. If FGB-2 eventually flies as the Commercial Space Module, it will actually perform that task, supplying large amounts of cargo and propellant to ISS before being used for other purposes. If this will be followed by any dedicated LTV flights remains to be seen.

Another measure announced after the January 1996 Gore-Chernomyrdin meeting to offset the loss of Progress-M2 was to increase the cargo capacity of the Progress spacecraft by 200 kg. This appears to have been part of an effort to gradually upgrade Progress by using ever more powerful versions of the Soyuz rocket under the Rus programme (some publications actually referred to the vehicle as Progress/Rus). It is known that in 1997-1998 RKK Energiya worked on an improved vehicle called Progress-MM that featured some of the same modifications as Soyuz-TMM. In 1999, simultaneously with the switch from Soyuz-TMM to TMS, those plans were abandoned in favour of a more modest modification named Progress-MS. The Draft Plan for this vehicle was approved by Semyonov in January 2000 and it is expected to make one or two flights *before* the first Soyuz-TMS to test the improved flight control system.

In early 1999 RKK Energiya also began studying an 11-12 tonne resupply ship to be orbited by the Yamal booster and sharing many design features with the ISS modules put forward by Energiya that same year. By the end of the year designers settled on a vehicle reminiscent of Progress-M2, but about 2 m shorter because Yamal had a smaller payload fairing than Zenit. Called Progress-M3, it had a total mass of about 11.5 tonnes and could deliver about 5.5 tonnes of cargo and propellant to ISS (Fig. 33). However, as Progress-M3 appeared on the drawing board, the focus shifted to a commercial version of Yamal ('Avrora') to be launched from Christmas Island and to the Enterprise module built jointly with Spacehab. Plans to launch Avrora from Baykonur's UKSS launch pad (used for the first Energiya launch in May 1987) have been abandoned, making it unlikely that Progress-M3 will ever be built [64].

A new iteration of Progress that *did* appear after the start of ISS assembly is Progress-M1. Remaining within the payload capacity of the standard Soyuz rocket, it carries more propellant to refuel ISS than Progress-M. Although the order to develop this vehicle was signed by RKK Energiya president Yuriy Semyonov on 15 February 1996, a Progress with an increased propellant supply was already mentioned in Russia's original ISS plans and it may have been developed independently from the Progress/Rus effort [65]. The increased propellant load is achieved by replacing the 'Rodnik' water tanks in the mid-section with four additional propellant tanks, giving a total of eight propellant tanks with a maximum of 1,700 kg of propellant (compared to 850 kg on Progress-M). Water (100 kg less than on Progress-M) is now stored in the cargo section. Twelve nitrogen and oxygen tanks were placed on the exterior of the craft around the intersection between the cargo and propellant compartments. Progress-M1 also has an improved computer (replacing the Argon-16), an autonomous navigation system using GPS and Glonass and an improved Kurs-MM rendezvous system, although the latter doesn't seem to have been used so far [66] (Fig. 34). The first two Progress-M1 cargo ships were launched to Mir in February and April 2000 and the first launch to ISS took place in August 2000.

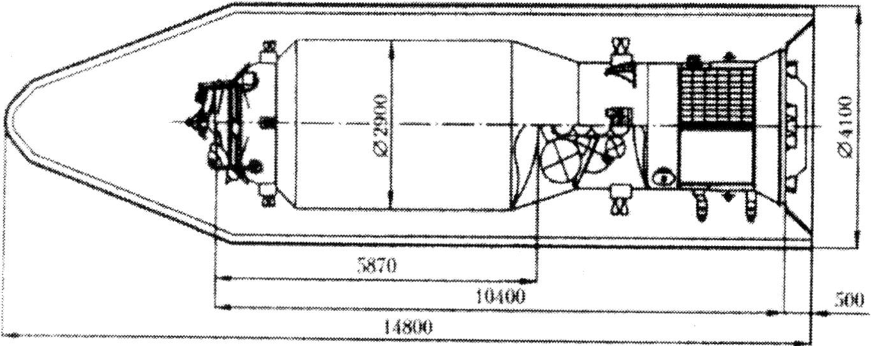

Fig. 33 Progress-M3.
        (source: RKK Energiya)

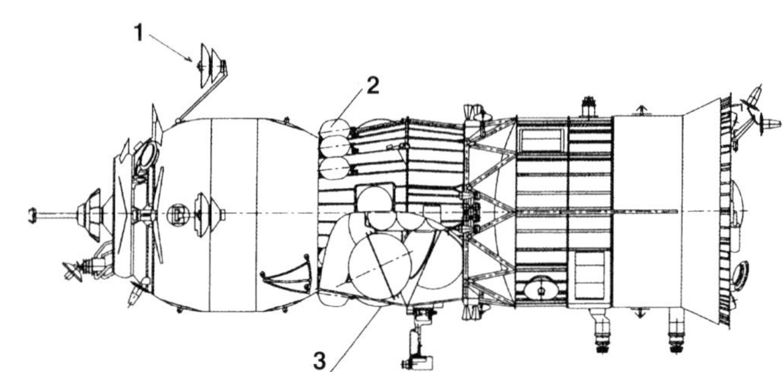

Fig. 34   Comparative views of Progr- ess-M (above) and Progress-M1.
(source : Novosti Kosmonavtiki)

Key: 1. Kurs-MM antenna;
2. Oxygen and oxygen-nitrogen tanks;
3. Propellant compartment with 8 propellant tanks.

Conclusion

Almost ten years after the International Space Station was conceived, the Russian segment is still a long way from reaching its final configuration. Russia's inability to complete its part of the station without Western financial support demonstrates that its domestic manned space programme would probably have been scaled back substantially or even cancelled without the space partnership it entered with the West in 1993. On the other hand, without critical Russian contributions such as the Service Module and the venerable Soyuz and Progress ferries, the ISS might never have been permanently manned as early as it did. Although largely built with Western money, it will probably also be Russian hardware that eventually will enable the station to be manned by the number of astronauts required to fully exploit its scientific potential.

Acknowledgments

The author would like to thank Konstantin Lantratov, who frequently writes about the ISS Russian segment for the Russian space magazine *Novosti Kosmonavtiki*. Without his contributions it would have been impossible to write this article.

References

1.   This section largely based on: K. Lantratov, "Zvezda: Road Into Space" (in Russian), *Novosti Kosmonavtiki*, 9/2000, pp. 3-7.

2.   For more background on KB Salyut and its involvement in space station work see: B. Hendrickx, "The Development of Mir and its Modules", in: R. Hall (ed.), *The History of Mir 1986-2000*, British Interplanetary Society, London, pp. 3-19, 2000.

3.   "Soviets Preparing Energia Booster/Buran-2 at Baikonur As Follow-On Mir-Station Is Canceled in Economic Crisis", *Aviation Week & Space Technology*, 22 April 1991, p.23.

4.   T. Furniss, "Module Replacement Will Extend Mir's Life", *Flight International*, 17-23 July 1991, p.16 ; A. Dolgikh, "Let Me Introduce You To Mir Stage Two" (in Russian), *Krasnaya Zvezda*, 2 August 1991 (translated by *JPRS Report/USSR Space*, 20 September 1991, pp.10-11).

5.   P. Langereux, "L'URSS a prévu d'édifier la station Mir 2 avec les navettes Bourane en 1996-2000", *Air & Cosmos*, 21-27 October 1991, pp. 44-45; L. van den Abeelen, "Mir and Buran to Become Integrated", *Spaceflight*, **34**, pp.293-294, 1992.

5   J. Lenorovitz, "Russia Redesigns Mir-2; Primary Module Underway", *Aviation Week & Space Technology*, 10 August 1992, pp.62-63; "The Mir-2 Longterm Orbital Station" (in Russian), *Novosti Kosmonavtiki*, 25/1992, pp. 16-19 ; C. Covault, "Russia Forges Ahead On Mir-2", *Aviation Week & Space Technology*, 15 March 1993, pp.26-27.

7.   V. Kirillov, "Progressive Progress" (in Russian), *Novosti Kosmonavtiki*, 4-5/1998, p.48.

8.   J. Lenorovitz, "Russia To Expand Role in Manned Space Flight", *Aviation Week & Space Technology*, 2 August 1993, pp. 62-63 ; K. Lantratov, "Ships for ISS" (in Russian), *Novosti Kosmonavtiki*, 7/1994, pp.23-25.

9.   J. Lenorovitz, "Proven Technology Is Cornerstone Of Mir-2", *Aviation Week & Space Technology*, 23 August 1993, pp.60-62; V. Filin et al., *S.P. Korolyov Space Corporation Energia: From First Satellite To Energia-Buran and Mir*, RKK Energiya, Moscow, p.120, 1994.

10.   J. Lenorovitz, "ESA Considers Further Ventures With Russia On Space Systems", *Aviation Week & Space Technology*, 7 September 1992, pp.36-37; "Germans Propose Russian Connection For Future MTFF", *Aviation Week & Space Technology*, 5 October 1992, p.27; I. Chornyy, "Lieber, Lieber, Amore, Amore..." (in Russian), *Novosti Kosmonavtiki*, 23-24/1998, pp.62-63.

11.   "ESA Seeks Space Co-operation With Russia", *Spaceflight*, **35**, pp.326-327, 1993.

12.   Yu. Semyonov (ed.), *Raketno-kosmicheskaya korporatsiya Energiya imeni S.P. Korolyova 1946-1996*, RKK Energiya, Moscow, p.496, 1996.

13.   B. Burrough, *Dragonfly: NASA and the Crisis Aboard Mir*, Harper Collins Publishers, New York, p.265, 1998.

14.   The letter is published in: Yu. Semyonov, *RKK Energiya...1946-*

1996, op. cit., p.497 ; Details of the assembly plan are from: A. Lawler, "Yeltsin Backs Merging Mir Program With Freedom Plan", *Space News*, 5-11 April 1993, p.1, 11.

15. A. Lawler, "Russians On Way To Help NASA Station", *Space News*, 12-18 April 1993, p.1, 20.

16. Yu. Semyonov, *RKK Energiya...1946-1996*, op. cit., p.498; A. Lawler, "Russia Eager To Join Space Station Team", *Space News*, 3-9 May 1993, p.4, 21; A. Lawler, "Russians Leave US, Set June Deadline", *Space News*, 10-16 May 1993, p. 1, 20 ; J. Lenorovitz, "Russian Cost Estimates Due On US Station Modules", *Aviation Week & Space Technology*, 17 May 1993, pp.33-34.

17. L. Tucci, "Team Bombarded by Redesign Ideas", *Space News*, 22-28 March 1993, p.1, 20.

18. A. Lawler, "June Decision Looms for Station Team", *Space News*, 26 April-2 May 1993, p.1, 20.

19. A. Lawler, "Russia Angered By U.S. Warning", *Space News*, 21-27 June 1993, p.1, 20; A. Lawler, "US, Russia Ink Trio of Co-operation Pacts", *Space News*, 26 July-1 August 1993, p.6; "US, Russia Settle Export Disagreement", *Aviation Week and Space Technology*, 26 July 1993, p.27.

20. J. Lenorovitz, "Russians Present Station Price List", *Aviation Week & Space Technology*, 25 October 1993, pp.68-69.

21. A. Lawler, "Chernomyrdin in Washington", *Space News*, 30 August-6 September 1993, p.4, 21.

22. The text of the pact is published in: "Joint Russian-American Commission on Energy and Space: Joint Statement on the Development of Co-operation in the Area of Space" (in Russian), *Novosti Kosmonavtiki*, 18/1993, pp.5-7; also see: P. Mann, "US, Russia Draft Historic Space Pact", *Aviation Week & Space Technology*, 6 September 1993, pp.22-23.

23. J. Asker, "NASA Details New Station Plans", *Aviation Week & Space Technology*, 13 September 1993, pp.20-21.

24. K. Lantratov, "Russia-USA. Into Space Together" (in Russian), *Novosti Kosmonavtiki*, 26/1993, pp.13-16.

25. Yu. Zhuravin, "What Is The FGB?" (in Russian), *Novosti Kosmonavtiki*, 1/1999, pp.2-3. According to this source the Russian Space Agency initially refused to include the FGB as an element of the Russian segment because of objections raised by RKK Energiya. It was not until NASA expressed its readiness to purchase the FGB, that the Russian Space Agency went along.

26. K. Lantratov, "Russia-US. Into Space Together", op. cit., p.17.

27. K. Lantratov, "Russia-US. Into Space Together", op. cit., p.18 ; L. Tucci, "Partnership Yields Sketchy Details On Savings", *Space News*, 29 November-5 December 1993, p.3.

28. K. Lantratov, "Russia-USA. New ISS Assembly Schedule" (in Russian), *Novosti Kosmonavtiki*, 3/1994, pp.32-37.

29. "ISS Assembly Sequence Changed" (in Russian), *Novosti Kosmonavtiki*, 14/1994, pp.47-48; I. Lisov, "ISS Assembly Schedule Changed" (in Russian), *Novosti Kosmonavtiki*, 20/1994, p.44.

30. NASA initially referred to it with the English acronym APDS, but later adapted the name to fit the Russian acronym, calling the system Androgynous Peripheral Assembly System.

31. Private correspondence with Konstantin Lantratov, 7 June 2002.

32. K. Lantratov, "International Space Station Alpha" (in Russian), supplement to *Novosti Kosmonavtiki*, June 1995, p.7, 18, 21-24; K. Lantratov, "New Russian Element of ISS Launched" (in Russian), *Novosti Kosmonavtiki*, 11/2001, p.7.

33. S. Valyayev, "New Proposals On ISS Assembly Being Prepared" (in Russian), *Novosti Kosmonavtiki*, 21/1995, p.49; O. Shinkovich, "New Scenario For Building the Station" (in Russian), *Novosti Kosmonavtiki*, 23/1995, p.70; "About A New Russian Plan for Assembling ISS" (in Russian), *Novosti Kosmonavtiki*, 24/1995, pp.52-53; W. Ferster and B. Iannotta, "NASA Mulling Alterations To the Space Station Plan", *Space News*, 4-10 December 1995, p.1, 21; P. de Selding, "Russia Seeks To Join Early Station Elements To Mir", *Space News*, 11-17 December 1995, p.4, 37; J. Asker and J. Anselmo, "Space Station Faces Russian Funding Crisis", *Aviation Week & Space Technology*, 11 December 1995, p.35; B. Iannotta, "Russian Module Damaged", *Space News*, 18-24 December 1995, p.4, 21.

34. Zarya was the name first proposed for the civilian DOS space stations. Just a few days before the launch of the first DOS, it was decided to name the station Salyut instead. However, it was launched with the name "Zarya" painted on the hull and on the nose fairing of the Proton rocket. Zarya was also the name of the Zenit-launched "Big Soyuz" (14F70) designed by NPO Energiya in the 1980s. Finally, Zarya was an unofficial name for the Baykonur cosmodrome in the late 1950s and early 1960s and it was also used as the call-sign of capcoms talking to orbiting cosmonauts in the early years of the Soviet manned space programme.

35. Main sources used for Zarya section: K. Lantratov, "The FGB Energy Block" (in Russian), *Novosti Kosmonavtiki*, 4/1994, pp.25-27; C. Covault, "Russian FGB 'Space Tug' Leads Station Alpha", *Aviation Week and Space Technology*, 4 September 1995, pp.48-51; V. Sorokin, "The FGB Will Be Modified" (in Russian), *Novosti Kosmonavtiki*, 7/1997, pp.48-50; Yu. Zhuravin, "What Is The FGB?" (in Russian), *Novosti Kosmonavtiki*, 1/1999, pp.2-7.

36. For more background on this see: J. Oberg, *Star-Crossed Orbits: Inside the US-Russian Space Alliance*, McGraw Hill, New York, pp.231-247, 2002.

37. The name Zvezda first emerged in the early 1960s for a space station designed by Korolyov's OKB-1 design bureau. Korolyov's engineers were reportedly inspired by the Russian SF novel *Zvezda KETs* written by Aleksandr Belyayev (KETs standing for "Konstantin Eduardovich Tsiolkovskiy"). This was the name of a space station described in the novel. Later the name was used for the following projects: a spaceplane proposed by the Tupolev bureau in the early 1960s; a manned military vehicle designed by Branch Nr. 3 of OKB-1 in 1965-1967, also known as 7K- VI and 11F72 ; an Almaz military space station with multiple docking ports designed by the Chelomey design bureau in the 1970s ; a lunar base proposed by NPO Energiya in the mid-1970s. For the origins of the name Zvezda see: T. Varfolomeyev, "Soviet Rocketry That Conquered Space: Early Space Station Projects by OKB-1, 1954-1965", *Spaceflight*, **42**, pp.384-385, 2000.

38. This section based on: K. Lantratov, "The Zvezda Service Module" (in Russian), *Novosti Kosmonavtiki*, 9/2000, pp.7-14.

39. V. Voronin, "The Universal Docking Module" (in Russian), *Novosti Kosmonavtiki*, 10/1998, p. 36.

40. This section based on: K. Lantratov, "International Space Station Alpha" (in Russian), supplement to *Novosti Kosmonavtiki*, June 1995, pp. 18-22.

41. I. Lisov and S. Golovkov, "6th Session of the Gore-Chernomyrdin Commission" (in Russian), *Novosti Kosmonavtiki*, 3/1996, pp.16-20; W. Ferster, "Russia Drops Plans For Ukraine's Zenit", *Space News*, 13-19 May 1996, p.3, 27.

42. I. Lisov, "Decisions of the ISS Control Board" (in Russian), *Novosti Kosmonavtiki*, 10/1997, pp.39-43; V. Sorokin, "Assembly of the ISS Russian Segment" (in Russian), *Novosti Kosmonavtiki*, 18-19/1997, pp.65-66; K. Lantratov, "New Russian ISS Element Launched" (in Russian), *Novosti Kosmonavtiki*, 11/2001, pp.7-8.

43. This section based on: "Statement on Russian-Ukrainian Co-operation" (in Russian), *Novosti Kosmonavtiki*, 11/1997, p.33; V. Voronin, "The Universal Docking Module" (in Russian), *Novosti Kosmonavtiki*, 10/1998, pp.36-37; V. Kirillov, "More Changes in the Russian Segment" (in Russian), *Novosti Kosmonavtiki*, 10/1998, pp.38-39; Yu. Zhuravin, "Ukrainian Module for ISS"/"News from the Russian Segment" (in Russian), *Novosti Kosmonavtiki*, 1/1999, p.26, 28; Yu. Zhuravin,

"Russian Segment of the ISS: Terra Incognita" (in Russian), *Novosti Kosmonavtiki*, 9/1999, p.63; V. Mokhov, "The Russian Segment of the ISS Becomes Ever More Commercial" (in Russian), *Novosti Kosmonavtiki*, 10/2000, p.18; Yu. Semyonov (ed.), *Raketno-kosmicheskaya korporatsiya Energiya imeni S.P. Korolyova na rubezhe dvukh vekov 1996-2001*, RKK Energiya, Moscow, pp.427,595, 2001.

44. Yu. Zhuravin, "Russian Segment of the ISS: Terra Incognita", *Novosti Kosmonavtiki*, 9/1999, p.62; Yu. Zhuravin, "FGB-2: Sold!" (in Russian), *Novosti Kosmonavtiki*, 9/2000, pp.20-21; Yu. Semyonov, *RKK Energiya...1996-2001*, op. cit. pp.411, 427, 440-442, 980.

45. B. Berger, "First Commercial Deal Penned for Station", *Space News*, 20 December 1999, p.3, 20; K. Lantratov, "Enterprise for ISS" (in Russian), *Novosti Kosmonavtiki*, 2/2000, p.62; Yu. Semyonov, *RKK Energiya...1996-2001*, op. cit. pp.507-508, 633-634.

46. Yu. Zhuravin, "FGB-2: Sold!" (in Russian), *Novosti Komonavtiki*, 9/2000, pp.20-21.

47. V. Mokhov, "The Russian Segment of ISS Is Becoming Ever More Commercial" (in Russian), *Novosti Kosmonavtiki*, 10/2000, p.18; Yu. Semyonov, *RKK Energiya...1996-2001*, op. cit. pp.508-509, 512, 514.

48. Yu. Semyonov, *RKK Energiya...1996-2001*, op. cit., pp.511-514, 634.

49. V. Mokhov, "Enterprise Legalised" (in Russian), *Novosti Kosmonavtiki*, 5/2001, p.30.

50. I. Afanasyev, "Agreement Between Boeing and Rosaviakosmos" (in Russian), *Novosti Kosmonavtiki*, 6/2001, pp.70-71; V. Mokhov, "Boeing and Khrunichev Will Build A Commercial Module After All" (in Russian), *Novosti Kosmonavtiki*, 7/2001, p.12.

51. Since there were no EVA exit hatches on Mir's Docking Module, it was cylindrical in shape with a maximum diameter of 2.2 m. Attached to either end of this cylinder was the forward part of a Soyuz orbital module with an APAS-89 docking port. In late 1998 there reportedly was talk of salvaging the Mir Docking Module for reuse on ISS, but the idea was rejected. See: "Reuse of Mir Module Deemed Too Expensive", *Space News*, 14-20 December 1998, p.2.

52. V.V. Favorskiy and I.V. Meshcheryakov, *Voenno-kosmicheskiye sily: kniga 2*, Izdatelstvo Sankt-Peterburgskoy tipografii, Moscow, p.131, 1998.

53. Some Western reports took over the Cyrillic spelling CO. Progress M-SO1 had a serial number in the Progress-M series (11F615A55 N° 301). No international designator was applied to Pirs itself, only to the combination of Pirs and the propulsion compartment (2001-041A). The propulsion compartment retained that designator after being separated from Pirs.

54. This section compiled from: "US Delegation Sees Adapter Soviets Will Use to Dock Buran, Mir", *Aviation Week and Space Technology*, 11 June 1990, p.49; Yu. Semyonov (ed.), *Mnogorazovyy orbitalnyy korabl Buran*, Mashinostroyeniye, Moscow, pp. 53-54, 1995 ; K. Lantratov, "A New Russian Element of ISS Launched" (in Russian), *Novosti Kosmonavtiki*, 11/2001, pp.6-10.

55. K. Lantratov, "Composition of the Russian ISS Segment Changed Again" (in Russian), *Novosti Kosmonavtiki*, 10/2001, pp. 22-23 ; V. Mokhov, "Agreement Signed on Finishing FGB-2"/"... And Will Enterprise Be Switched To The Shuttle?" (in Russian), *Novosti Kosmonavtiki*, 2/2002, pp.18-19; V. Mokhov, "The New Look of the ISS Russian Segment", *Novosti Kosmonavtiki*, 3/2002, p.14; personal correspondence with K. Lantratov, 7 June 2002, 13 August 2002.

56. V. Lukiashchenko et. al., "MAKOS-T – A New Spacecraft for Conducting Experiments in Microgravity", *Russian Space Bulletin*, 4/1996, pp.13-15; K. Lantratov, "Rosaviakosmos Orders MAKOS" (in Russian), *Novosti Kosmonavtiki*, 12/1999, p.52; Yu. Zaytsev, "New Russian Projects" (in Russian), *Novosti Kosmonavtiki*, 4/2002, pp.16-17; Yu. Semyonov, *RKK Energiya...1996-2001*, op. cit., pp.534-536, 581.

57. This section based on: C. Covault, "Mir Cosmonauts Prepare for Reentry As NASA Holds Soyuz Talks in Moscow", *Aviation Week & Space Technology*, 23 March 1992, p.24 ; B. Henderson, "Lockheed Strikes Deal With Soyuz Builder", *Aviation Week & Space Technology*, 14 September 1992, p.83; "Will Soyuz Become A Rescue Ship For Freedom?" (in Russian), *Novosti Kosmonavtiki*, 24/1992, pp.17-19; K. Lantratov, "Ships for ISS" (in Russian), *Novosti Kosmonavtiki*, 6/1994, pp.19-21 – 7/1994, pp. 22-23 ; Yu. Semyonov, *RKK Energiya...1996-2001*, op. cit., pp.494-495, 517-520.

58. Yu. Semyonov, *RKK Energiya...1946-1996*, op. cit., p.521.

59. FG stands for "forsunochnaya golovka", the Russian word for fuel injector. The first and second stage engines have been redesignated: the engines of the four strap-on boosters are called RD-107A or 14D22 (as compared to RD-107/11D512 on the 'old' Soyuz) and the core stage engine is named RD-108A or 14D21 (as compared to RD-108/11D511). The improved fuel injectors are one of the modifications under the so-called "Rus" programme that will eventually lead to the development of the Soyuz-2 rocket. See: I. Afanasyev, "Progress M1-6 Launched" (in Russian), *Novosti Kosmonavtiki*, 7/2001, p.9.

60. K. Rusakov, "Soyuz TMA – Ship for ISS" (in Russian), *Novosti Kosmonavtiki*, 15-16/1998, pp.54-56; S. Shamsutdinov, "The Soyuz TMA Ship" (in Russian), *Novosti Kosmonavtiki*, 17-18/1998, p.53; Yu. Semyonov, *RKK Energiya...1996-2001*, op. cit., pp.356, 600-628.

61. Details of Soyuz-TMS and TMM are from: Yu. Semyonov, *RKK Energiya...1996-2001*, op. cit., pp.628-631.

62. A. Zak, "Russia and NASA Square Off Over the ISS", article on space.com website, 7 November 2001.

63. Yu. Zhuravin, "Russian ISS Segment: Terra Incognita" (in Russian), *Novosti Kosmonavtiki*, 9/1999, pp.62-63.

64. Details of Progress-MM, MS and M-3 are from: Yu. Semyonov, *RKK Energiya...1996-2001*, op. cit., pp.631-634, 730-731; O. Urusov, "Avrora Will Not Fly From Baykonur" (in Russian), *Novosti Kosmonavtiki*, 1/2002, p.48.

65. A "tanker" version of the standard Progress-M is mentioned in: K. Lantratov, "Ships for ISS" (in Russian), *Novosti Kosmonavtiki*, 7/1994, p.23.

66. K. Rusakov, "The Transport Cargo Ship Progress M1" (in Russian), *Novosti Kosmonavtiki*, 4/2000, p.16; Yu. Semyonov, *RKK Energiya...1996-2001*, op. cit., p.579.

* * *

ISS American Elements On-Orbit Through December 2001

ROELOF SCHUILING

Pressurized Elements

The pressurized elements provide work and living spaces. Pressurized structures are either primary or secondary. Primary structures are those designed to maintain the structural integrity of the pressurized modules while secondary structures are designed to transfer loads to primary structures.

Primary structures include such as ring frames, longeron-stiffened pressure shells, windows, or integrated trunnions. The typical design for a pressurized element consists of ring frames with longerons and shell panels. The longerons are used to increase the stiffness and load-carrying capability of the shell panels and the shell panels form the walls of the pressurized element. The integrated trunnions are used to mount the pressurized elements in the Space Shuttle orbiter during the trip to the International Space Station (ISS) and they lock into retention latches in the orbiter's payload bay.

The secondary structure may be both internal and external to the pressurized element and includes such items as crew and payload translation aids, equipment support, and debris shielding. Internal secondary structures also include rack attachment structures and passages for electrical and thermal cabling.

The pressurized element debris protection is by external Micro-Meteoroid Orbital Debris (MM/OD) shielding. This passive system consists of a 1.27 mm thick sheet of aluminum separated from the pressure shell by a 1-1.6 mm gap. The debris shielding shocks the orbital debris and breaks it into small fragments which spread the energy over a much larger area, thereby causing much less damage. MM/OD shielding for windows is provided by use of window shutters.

The Common Berthing Mechanism (CBM) connects one pressurized element to another on the U. S. segment of the ISS. The CBM has an active and passive half. The active half contains a structural ring, capture latches, alignment guides, powered bolts, and controller panel assemblies. It is the only half connected to power and data lines. The passive half has a structural ring, capture latch fittings, alignment guides, and nuts. During the installation of a pressurized element a robotic arm moves the element with the passive half into the capture envelope of the active half, after which the latching process takes place.

Unity; The First American International Space Station Element

At 3:35 on the morning of Friday, December 4, 1998 the Space Shuttle Endeavour lit up the dark Florida night as it was launched from the Kennedy Space Center. It was carrying the first of the American components for the ISS. The STS-88 (ISS Mission 2A) mission goal was the delivery of Node One, named "Unity", to mate with the Russian Zarya ("Sunrise" in Russian) control module. The name "Unity" honored the spirit of international cooperation and joint achievement reflected worldwide in the work of those building the ISS. The Unity Node joined the modules of former Cold War adversaries Russia and the United States. The Zarya had been launched a few weeks earlier from the Baikonour Cosmodrome on November 20, 1998.

Unity is an aluminum cylindrical pressurized module with six berthing ports. One port is located at either end and four were located at 90-degree spacing intervals around the center portion of the Unity. In addition to its connection with Zarya, it would provide habitable pressurized space and a passageway to other ISS elements. Unity connected to the U. S. Laboratory module, an airlock, the Z1 truss, and berthing capability was reserved for future connection to a U. S. habitation module and a windowed cupola . The Unity has a length of 5.5 meters and a diameter of 4.6 meters. It contains over 50,000 mechanical items, 216 fluid and gas lines, 121 internal and external cables and uses over 9.6 kilometers of wire. Unity and two Pressurized Mating Adapters weighed 11,600 kilograms at launch and had a combined length of almost 11 meters.

A close up view of Node 1 (Unity) in its work stand in the Space Station Processing Facility. (NASA)

The ports at either end were attached to Pressurized Mating Adapters (PMA)s. The PMAs are truncated conical pressurized units designed to mate with connecting ports on the Shuttle orbiter and other ISS elements. Those attached to Unity on the STS-88 mission were termed PMA-1 and PMA-2.

PMA-1, attached to the aft end of Unity, was the mating unit that was connected to Zarya during the STS-88 mission on December 6, 1998. PMA-2, at the forward end, is used to mate with the Space Shuttle orbiter and provide entry for Shuttle crews to the ISS. PMA-1 was planned to be permanently attached to Unity and will serve as the connection to Zarya while PMA-2 was later removed to make way for the Destiny laboratory module during the STS-98 (ISS Mission 5A) mission in February of 2001. After the removal and temporary placement of PMA-2 on one of the station trusses PMA-2 was then attached to the Destiny laboratory on the opposite end from Unity. There it serves as a docking port for Shuttle orbiters

The Boeing Company at the Marshall Space Flight Center in Huntsville, Alabama built the Unity Node. All preflight dynamic and structural testing was also done at the Marshall Center. Boeing also built the PMAs attached to Unity. Unity was shipped to the Kennedy Space Center on June 22, 1997 and arrived at the Florida launch site on the following day. PMA-1 was mated to Unity about 10 days following the arrival while PMA-2 was mated in early 1998.

During the STS-92 (ISS Mission 3A) Space Shuttle mission in October of 2000 a third PMA, PMA-3, was mounted to the Unity Node to provide the capability for the Space Shuttle orbiter to dock with the ISS and then, while docked, remove the PMA-2. The PMA-2 was temporarily stored on the Z1 truss so that the Destiny laboratory could be connected to the Unity Node on the STS-98 mission of February 2001. The PMA-2 was later moved to the front of the Destiny laboratory to serve as a Space Shuttle orbiter docking port.

PMAs included mechanical interfaces, spacewalk hardware and thermal control heater equipment, an electrical power subsystem, and Command and Data Handling passthroughs. The PMAs provide a hard line 1553 data bus connection between the orbiter and Unity via X-Connectors on the Androgynous Peripheral Attachment System. This system interfaces with the Shuttle orbiter's Orbiter Docking System. Umbilical connections between the PMA and Unity complete the hard line path. This path allows orbiter interface units and portable computer systems to communicate on the ISS-orbiter buses. The PMA is equipped with Extravehicular Activity (EVA) aids including a Portable Foot Restraint, a top mounted Worksite Interface Fixture, two Flight Releasable Grapple Fixtures, camera and laser targets, Space Vision Targets, handholds and handrails. The Flight Releasable Grapple Fixture provides the standard mechanical interface between the Shuttle's robotic arm and payloads. It can be released during an EVA by rotating two release rods enabling the grapple shaft to be removed.

PMAs –2 and -3 have two sets of four red light emitting diodes that tell the orbiter crew the status of the ISS attitude control system. The crew can see the diodes through the overhead windows on the orbiter Aft Flight Deck.

Node 1 (Unity) being connected to the Zarya module.   (NASA)

Each set is controlled by a separate Unity Node Multiplexer-Demultiplexer for one-fault tolerance during arrival or departure of the orbiter. Free drift is indicated by the two sets of Light Emitting Diodes flashing on or off at a 5-hertz rate. A steady light indicates every other state on.

The PMA is a truncated conical shell with a 0.6-meter axial offset in the diameters between the end rings. They are ring-stiffened shell structures machined from 2219 aluminum alloy roll ring forgings welded together.

Endeavour's STS-88 mission was successful in bringing Unity and Zarya together. On December 5, 1998 astronaut Nancy Currie used the Space Shuttle orbiter's robotic arm to lift Unity and the two PMAs out of the orbiter payload bay and mounted PMA-2 to the Orbiter Docking System. As Unity projected up out of the payload bay Endeavour proceeded to close with Zarya, arriving on the evening of December 6th. Currie then used the robotic arm to grapple Zarya and bring it down to mate with the PMA-1 on the end of Unity. Following an Extravehicular Activity spacewalk by astronauts Jerry Ross and Jim Newman to hook up data and power cables between the Russian and American elements, Unity was activated on December 7th. Following additional tests and preparations the STS-88 crew entered through the Orbiter Docking System and into Unity by way of PMA-1. They later passed through PMA-2 and into the pressurized Zarya module. After years of preparation and planning the ISS was now operating with a human crew.

Two additional Space Shuttle flights visited the growing ISS before another set of American hardware joined the station. On May 27, 1999 the orbiter Discovery was launched on STS-96 (ISS Mission 2A.1) to bring supplies to the ISS. Following the July 12, 2000 launch of the Russian Zvezda service module to join with Zarya and Unity, Atlantis' STS-106 (ISS Mission 2A.2b) mission lifted off from Florida on September 8, 2000. The STS-106 mission featured EVA operations to install equipment and connect cables between the station components as well as transfer supplies into the ISS in preparation for upcoming station operations.

The Z1 Zenith Truss

The Z1 truss was an early exterior structure allowing the temporary mounting of the first U. S. solar arrays on Unity to provide early electrical power. A month after the STS-106 flight the Space Shuttle orbiter Discovery brought the next American components to the ISS on the STS-92 mission. On October 11, 2000 STS-92 was launched and brought the first of the ISS truss segments; the Z1 truss, as well as the third Pressurized Mating Adapter; the PMA-3, to the ISS. On the third day of the mission, October 14th, Japanese Astronaut Koichi Wakata transferred the Zenith Z1 truss from Discovery's payload bay using the orbiter's robotic arm. He deftly raised the truss out of the bay and berthed it to the zenith port of the Unity Node One. On the 16th, the PMA-3, also flying on STS-92, was transferred from Discovery's payload bay to the nadir port of Unity. There it served as a docking port for the Space Shuttle orbiter while PMA-2 was unavailable due to the need to add the Destiny laboratory to the Unity Node later in the assembly sequence.

Part of the Z1 element was the Motion Control Subsystem consisting of the Control Moment Gyros and their assemblies. The four Control Moment Gyros, which control the attitude of the ISS, have a spherical momentum storage capability of 18,981.5 Joules per second, the scalar sum of the individual Control Moment Gyro wheel moments. The momentum stored in the gyro system at any given time is equal to the vector sum of the individual Control Moment Gyro momentum vectors. In order to maintain the ISS in a desired attitude the Control Moment Gyro system must cancel, or absorb, the momentum generated by the disturbance torques acting on the ISS. If the average disturbance torque is not zero then the resulting Control Moment Gyro output torques is also not zero and momentum builds up in the Control Moment Gyro system. When the system becomes saturated it is unable to generate the torques required to cancel the disturbance torques. The result is then a loss of attitude control. The Control Moment Gyro system saturates when momentum vectors have become parallel and only momentum vectors change. In this case control torques perpendicular to this parallel line are possible and controllability about the parallel line are lost.

A Control Moment Gyro consists of a large flat wheel rotating at a constant 6,600 revolutions per minute and developing an angular momentum of 4,745.36 Joules per second about the spin axis. This rotating wheel is mounted in a two-degree-of-freedom gimbal that allows pointing the spin axis – the momentum vector – of the wheel in any direction. At least two Control Moment Gyros are required to provide attitude control. The Control Moment Gyro generates an output reaction torque that is applied to the ISS by inertially changing the direction of the wheel momentum. The Control Moment Gyro's output torque has two components, one proportional to the rate of change of the Control Moment Gyro gimbals and a second proportional to the inertial body rate of the ISS as sensed at the Control Moment Gyro base. Because the momentum along the direction of the spin axis is fixed, the output torque is constrained to lie in the plane of the wheel – thus one Control Moment Gyro cannot provide the three-axis torque required to control the ISS attitude.

The Control Moment Gyros were not activated until the STS-98 (ISS Mission 5A) and each Control Moment Gyro had a thermostatically controlled survival heater to keep it within thermal limits before activation. The 120-watt heaters have an operating temperature range of minus 41.11 to minus 37.22 degrees Celsius.

The Z1 Integrated Truss Segment contains both S-Band and Ku-Band communications systems. The S-Band system consists of two redundant strings with three on-orbit replaceable units and two antennas, the Baseband Signal Processor, the Tracking and Data Relay Satellite System (TDRSS) transponder, and the antenna RF group. The antenna group included both a low-gain and high-gain directional antenna.

The Radio Frequency Group amplifies and filters the radio signals and receives and radiates the signal, providing the interface with free space. It also controls antenna switching and pointing. The unit antennas contain a high-gain antenna that can support the high-rate data link but it requires TDRSS pointing updates that are generated by the guidance, navigation, and control pointing function. The updates were not available at the time of the STS-92 mission, however, as they were not scheduled to be available until the guidance, navigation, and control multiplexer-demultiplexer was activated after the Destiny laboratory module was added to the ISS on STS-98. The low-gain antenna is fixed in position and needs no pointing data.

The Baseband Signal Processor is the core of the S-Band system and provides data and voice processing for downlink and uplink. A voice line connects directly to the processor to provide voice communications. The TDRSS transponder, which was not present until mission STS-97 (ISS Mission 4A), receives downlink signals from the processor and modulates the RF carrier for transmission to the ground. The transponder receives the uplink information from the Radio Frequency group, demodulates the signal, and routes it to the processor for information distribution. The downlink output of the Baseband Signal Processor is a constant-rate data stream of either 192 kbps or 12 kbps, as required. The processor inputs a digital bit stream from the TDRSS transponder at either 72 or 6 kbps, dependant upon whether the processor is configured for a high or low data rate. The data is processed by decryption, Consultative Committee for Space Data Systems header validation, and demultiplexing functions. If uplink audio channel data is present, it is expanded to 192 kbps per channel for interface to the Audio Interface Unit and routed to an output port. The uplink core data channel packets are passed to a 1553 interface for transfer to the command and control processor.

The Ku-Band communications system is the primary return link from the ISS to Earth for video and payload data transmitted in digital format. The Ku-Band provides a 50 mbps fixed-rate downlink with up to four video signals or up to 43 mbps high-rate data with 7 mbps overhead. A communication outage recorder records payload data in the zone of exclusion and during structural signal blockages. There are 12 logic channels – 4 video and 8 payload – in the Ku-Band downlink. Both the Ku-Band system and the S-Band systems do not inspect the data passing through them. The ISS transmits at a constant rate of 50 mbps, of which 43 mbps is available to the user. The video channels can be configured for downlinking full-motion or stop-action video. The later consists of skipping video frames. Normally the configuration is set by the ground and commanded through the S-Band uplink to the ISS Command and Data Handling system which routes commands through the 1553 local bus to the Ku-Band system.

The Ku-Band system contains a single string of four avionics units, Video Baseband Signal Processor, High Rate Frame Multiplexer, the High Rate Modem, the Transmit/Receive Controller, plus an erectable and steerable antenna. Video signals must be processed by the Video Baseband Signal Processor before they are routed to the High-Rate Frame Multiplexer for interleaving with the high-rate data. The high-rate data comes directly from the payload patch to the High-Rate Frame Multiplexer for processing and interleaving with the video signal. The High-Rate Frame Multiplexer builds the downlink frames and routes the data stream to the High-Rate Multiplexer for modulation on an S-Band interface frequency and routing to the Transmit/Receive Controller for translation to the Ku-Band frequency. The data stream is power amplified and routed to the steerable antenna for transmission through the TDRSS to the ground. An uplink carrier is received and processed by the Transmit/Receive Controller to maintain antenna autotracking only.

Video channels are routed to the Video Baseband Signal Processor for processing. The processor has 4 audio and 4 video input channels and 4 video plus audio output channels. Connections are provided for power and the 1553 bus and the interface data out. The 4 video plus audio output channels are routed to the High-Rate Frame Multiplexer. The Video Baseband Signal Processor performs audio and video analog-to-digital format conversions, packet formatting, built-in-test, and on-orbit replacement unit control functions. The processor converts incoming analog signals to digital, formats them for selected mode of operation, and forwards them to the High-Rate Frame Multiplexer for interleaving with high-rate payload data.

The electrical power subsystem has several power distribution components on the Z1 truss assembly. These include two Initialization Diode Assemblies, two secondary power distribution assemblies, two Plasma Contactor Units, two DC-to-DC Converter Units, and two patch panels. The Initialization Diode Assemblies provide diode-protected power from the Space Shuttle Assembly Power Conversion Unit to the P6 truss for initialization on mission STS-97 (ISS Mission 4A) and also provided a connection path from the P6 truss to the Destiny laboratory DC-to-DC Converter Units from the STS-98 - Destiny's arrival until the arrival of the P4 truss on ISS Mission 12A. The Secondary Power Distribution Assemblies included Remote Power Control Modules, power and data connections (Central Utility Rail), and a cold plate. They control, protect, and isolate secondary distribution lines, These distribution assemblies accept secondary power from the DC-to-DC Converter Units and distribute this power to the Remote Power Distribution Assembly or downstream loads while the cold plates transfer heat from the Remote Power Controller Modules by use of radiant fins.

The Central Utility Rail provides power connections from the DC-to-DC Converter Unit power feed to the Remote Power Controller Modules and data connections by way of redundant 1553B data busses. The Remote Power Controller Modules are electronic switches that control, protect, and isolate secondary distribution lines. There are six types of these controller modules, which vary in rated output current, number of switches per module, and trip function. The Plasma Contactor Units emit electrons through self-generated plasma and are self-regulating. They control voltage between the space plasma and the space station structure. The units mounted on the Z1 truss maintain the structural potential of the ISS within 40 volts of the plasma potential. A central element of each is their hollow cathode assembly, which emits up to 10 amps of electron current to the ambient plasma. They actively emit when the ISS is exposed to sunlight. Without them the ISS structural potentials could reach values of – 150 volts.

Two DC-to-DC Converter Units, converting power from primary (115 to 173 volts, DC) to secondary (123 to 126 volts DC) were attached to the starboard side of the Z1 during STS-92, however they were not activated until STS-97 when the P6 truss power came on line. The two patch panels on the port side of the Z1 truss allow the Z1 input power source to be changed. The launch configuration provided for Russian power, however during STS-97 they were reconfigured to allow for U.S. power from the P6 truss.

The Z1 truss included mechanical interfaces between the Z1 and the P6 truss on Z1's Zenith face, between Z1 and Node One on Z1's nadir face, between Z1 and PMA-2 on Z1's forward face for temporary stowage of the PMA-2 by way of a manual berthing mechanism, and a cable tray from Z1's forward face to the U. S. Destiny laboratory which arrived on ISS Mission 5A

The Z1 truss also included several EVA aids including two tool storage devices, 22 Worksite Interface Sockets, 1 flight-releasable grapple fixture, 11 trusses, 2 tray launch restraints, handholds, and handrails. The Z1 truss structure and shell are aluminum 219-T851, the keel pin elements are INCO 718 and the Ku-Band antenna boom elements are steel. A solid plate beneath the Control Moment Gyro face provides additional protection from orbital debris and micrometeorites. The Z1 Thermal Control System included accumulators, ammonia, and 12 quick disconnects. The accumulators charge the Thermal Control System with ammonia on-orbit, accommodate expansion in the fluid, and maintain the system's operating pressure. The quick disconnects provide for the connection of ammonia transfer lines between the P6 and Z1 truss and between the Z1 and the U.S. Destiny laboratory

PMA-3 arrived at the Kennedy Space Center's Space Station processing facility from the Boeing Huntington Beach, California factory on February 20, 1998 while the Z1 truss arrived at the Florida launch site on February 17, 1998. The addition of the 1,361 kilogram PMA-3, together with the 8,100-kilogram Z1 truss to the ISS resulted in the space station weighing approximately 72,575 kilograms following the STS-92 Shuttle mission.

The P6 Integrated Truss Structure

Providing early power to the ISS orbital elements was the next step in the U. S. space station element buildup. On November 30, 2000 the Space Shuttle Endeavour was launched on mission STS-97 to bring the P6 Integrated Truss Structure to the ISS. (The designation "P" or "S" refers to the truss's position being port or starboard on the main truss when completed) The P6 truss structure would ultimately be the most distant truss element on the one side of the completed ISS trusswork backbone. On December 3, 2000 astronaut Marc Garneau used the Endeavour orbiter's robot arm to place the P6 Integrated Truss Structure on the Z1 truss. The P6 extended upward from the Z1 such that when the solar arrays were opened they extended 90 degrees from the Unity Node's longitudinal axis.

The P6 would be removed later in the ISS assembly sequence and positioned at the port end of the ISS's truss backbone with the P1 truss element the closest port truss to the ISS pressurized modules. The S0 segment would be the central truss element and connect to the P1 truss on one side and the S1 segment on the starboard side. Five truss elements in final form would make up the starboard ISS truss segments. Five additional would form the port side with the S0 as the center element. Four of the ISS truss segments would ultimately carry photovoltaic solar array sets. Eight arrays in four sets of two will provide electrical power for the ISS. The STS-97's P6 Integrated Truss Structure consists of three major components including the Photovoltaic Array Assembly, the Integrated Equipment Assembly, and the Long Spacer. Together they weighed approximately 7,900 kilograms.

The Photovoltaic Array Assembly consists of two major elements. The Solar Array Assembly contains two Solar Array Wings; each made up of two panels, deployed in opposite directions and connected to a mast that was folded during launch and before removal from Endeavour's payload bay. The Beta Gimbal Assembly consists of the mast canister which housed the folded mast; the Bearing, Motor and Roll Ring module used to rotate the Solar Array Wing and transfer power; the Electronic Control Unit used to control the Beta Gimbal Assembly motor and mast rotation; and the Sequential Shunt Unit used to coarsely regulate the Solar Array Wing output voltage and route power to the Integrated Equipment Assembly.

The fully deployed Solar Array Wings were 35.02 meters long and 11.6 meters across. The two wings, deployed in opposite directions, spanned over 73 meters. Each wing weighs 1,088 kilograms and uses nearly 33,000 solar array cells. Each of the cells measures 8 centimeters square and contains 4,100 diodes. Each of the Solar Array Wings is capable of generating approximately 31 kilowatts of direct current. The two Solar Array Wings on the P6 generate enough power to meet the needs of 30 average homes without air conditioning.

The Beta Gimbal Assembly measures 2.4 x 2.4 x 0.6 meters and provides the structural link with the Integrated Electronics Assembly. Its functions include deploying and retracting the Solar Array Wing and rotating the wing about the wing's longitudinal axis. The motor used to rotate, deploy, and retract the solar arrays is a three-phase 200 watt DC stepper motor. The system has a rotational pointing accuracy of +/- 1 degree and a maximum rotational rate of +/- 200 degrees per minute. Rotational rate is controlled by modulation of the input DC power. The Roll Rings allow the transfer of power from the rotating (Mast Canister) to the stationary (Integrated Equipment Assembly) side and can transfer up to 35 kilowatts. The Electronic Control Unit controls the motor that rotates, deploys, and retracts the Solar Array Wings. It operates off of 120 volts DC and receives power from the DC-to-DC Converter Unit in the Integrated Equipment Assembly. Commanding the 0.6 x 0.3 x 0.3 meter unit is also from the Integrated Equipment Assembly. The Sequential Shunt Unit coarsely regulates power collected during periods of insolation during sun-pointing periods. A sequence of 82 separate power lines lead from the solar array to the shunt unit and shunting, or controlling, the output of each regulates the power transferred. The regulated voltage set point is controlled by a computer in the Integrated Equipment Assembly and is normally set to 140 volts. The Sequential Shunt Unit has an over voltage protection feature to maintain the output voltage below 200 volts DC maximum for all operating conditions. This power is then routed through the Bearing, Motor, and Roll Ring Module to the Direct Current Switching Unit in the equipment assembly. The Sequential Shunt Unit measures 0.8 x 0.5 x 0.33 meters in size and weighs 83 kilograms.

The second major P6 Integrated Truss Structure component is the 4.87 x 4.87 x 4.87 meter Integrated Equipment Assembly. Measuring almost 7,700 kilograms, it is designed to condition and store the electrical power collected by the photovoltaic arrays. It integrates the energy storage subsystem, the electrical equipment, the thermal control system, structural framework, and the Solar Array Rotary Joint and consists of three elements.

The first element is the power system electronics consisting of the Direct Current Switching Unit used for primary power distribution; the DC-to-DC Converter Unit used to produce regulated secondary power; the Battery Charge/Discharge Unit used to control the charging and discharging of the storage batteries; and the storage batteries themselves.

The Photovoltaic Thermal Control System consists of the coldplate subassembly used to transfer heat from an electronic box to the coolant; the Pump Flow Control System used to pump and control the ammonia coolant; and the Photovoltaic Radiator used to dissipate heat into deep space.

The third element is the set of computers used to control the P6 module.

The power system is divided into two independent and nearly identical channels. Each channel is capable of control, storage, and distribution of power to the ISS. Power received from each Photovoltaic Array Assembly is fed directly into the appropriate Direct Current Switching Unit. The switching unit is a high power, multi-path remotely controlled unit used for primary and secondary power distribution, protection, and fault isolation within the Integrated Equipment Assembly. It also distributes primary power to the ISS. During periods of insolation the switching unit routes primary power directly to the ISS from array assembly and also routes power to the power storage system for battery charging. During periods of eclipse the switching unit routes power from the power storage system to the ISS. It measures 0.8 x 1.02 x 0.3 meters in size and weighs 106 kilograms.

Primary power from the Direct Current Switching Unit is also distributed to the DC-to-DC Converter Unit. This unit conditions the coarsely regulated power from the Photovoltaic Array Assemblies to 123 +/- 2 volts DC. It has a maximum power output of 6.25 kilowatts. This power is used for all P6 operations employing secondary power. Power from the switching unit is also distributed to the three power storage systems located within each channel of the Integrated Equipment Assembly. The power storage system consists of a Battery Charge/Discharge Unit and two battery assemblies.

The Battery Charge/Discharge Unit serves a dual function of charging batteries during solar collection periods and providing conditioned battery power to the primary power busses during eclipse periods. The unit has a battery charging capability of 8.4 kilowatts and a discharge capability of 6.6 kilowatts. It also includes provisions for battery status monitoring and protection from power circuit faults. Commanding the unit is by way of the Integrated Equipment Assembly computer.

Each battery assembly consists of 38 lightweight nickel-hydrogen cells and associated electrical and mechanical equipment. Two battery assemblies connected in series are capable of storing a total of 8.3 kilowatts of power. This power is fed to the ISS via the Battery Charge/Discharge Unit and the Direct Current Switching Unit respectively. The battery design life is 6.5 years and can exceed 38,000 charge/discharge cycles at 35% depth of discharge. Each battery measures 1.02 x 0.9 x 0.46 meters and weighs 170 kilograms.

In order to maintain safe operating temperatures for the electronics, they are conditioned by the Photovoltaic Thermal Control System. This system consists of ammonia coolant, eight coldplates, two Pump Flow Control Systems, and one Photovoltaic Radiator.

The coldplate subassemblies are an integral part of the structural framework. Heat is transferred from the electronic boxes to the coldplates via fine interweaving fins located on both the coldplate and the electronic boxes. The fins add lateral stiffness in addition to increasing the available heat transfer area.

The Pump Flow Control Systems are the heart of the thermal system. They consist of all the pumping capacity, valves and controls required to pump the heat transfer fluid to the heat exchanges and radiator, and regulate the temperature of the ammonia coolant. The system is designed to dissipate 6,000 watts of heat per orbit and is commanded by the Integrated Equipment Assembly computer. Each consumes 275 watts during normal operations and measures 1.02 x 0.74 x 0.48 meters and weighs 106.5 kilograms.

The radiator was deployed on-orbit and is comprised of two separate flow paths through seven panels. Each path is independent and connected to one of the two Pump Flow Control Systems on the Integrated Equipment Assembly. The 725-kilogram radiator can reject up to 14 kilowatts of heat into deep space

The third major component of the P6 Integrated Truss Structure is the Long Spacer. Measuring 8.53 x 4.88 x 4.88 meters, the Long Spacer performs two basic functions. The first is to physically separate the P6 solar arrays from the adjacent arrays following the relocation of the P6 on a later assembly mission, and the second is to provide temporary cooling for the U. S. Destiny laboratory module. Later, Destiny will be cooled by the main Heat Rejection System after its activation on a later assembly flight.

The Long Spacer thermal system is the Early External Active Thermal Control System and is similar to the Integrated Equipment Assembly's Photovoltaic Thermal Control System with a few differences. The Long Spacer's system employs two separate cooling system loops and has two high-power Pump Flow Control Systems with ach feeding into its own Photovoltaic Radiator. It also has external ammonia accumulators used for pressure control and heaters to prevent freezing.

The Pump Flow Control System's control valve regulates the ammonia flow though the radiator in response to the Destiny laboratory heat exchanger temperature. The system is designed to reject 14,000 watts of heat per orbit and is commanded by the Pump Flow Control System under the control of the Integrated Equipment Assembly computer. Each of the two Pump Flow Control Systems consumes 350 watts during normal operations and the heaters consume 115 watts

Destiny Laboratory Module

The arrival of the U.S. Laboratory, named "Destiny", at the ISS was the next step in the space station assembly. The laboratory was launched aboard the STS-98 Space Shuttle orbiter Atlantis on February 7, 2001. With Atlantis docked to the bottom port of Unity on February 10, 2001 astronaut Marsha Ivins used the Atlantis' robot arm to remove the PMA-2 from the front port of the Unity Node and placed the PMA-2 temporarily on the Z1 truss. She then removed Destiny from the payload bay and mated the laboratory with the front port of the Unity. On February 12th the PMA-2 was transferred from its temporary Z1 location to the front of Destiny where it serves as a Shuttle orbiter docking port.

The Destiny laboratory serves as the centerpiece of scientific research aboard the ISS. And also serves as command and control center for the entire complex. The Boeing Company began construction of the 14, 500 kilogram laboratory at the Marshall Space Flight Center in Huntsville, Alabama in 1995. The laboratory was shipped to the Kennedy Space Center in 1998. On December 1, 1998 it was named "Destiny".

Destiny is constructed of three aluminum cylindrical sections 4.4 meters in diameter with a total cylindrical length of 7.7 meters. Two end-cones, containing the hatches through which astronauts enter and leave the laboratory bring the total exterior length to 8.8. meters. The laboratory itself weighs 14,514 kilograms while its average operational on-orbit weight comes to 26,771 kilograms. A nadir-mounted Widow Observational Research Facility is positioned so as to allow Earth observation from the ISS's usual attitude. Experiments within the

In the Space Station Processing Facility work continues on the U.S. Lab module, Destiny. (NASA)

laboratory have access to power, communications, cooling, vacuum, exhaust, gaseous nitrogen and measurement resources. The internal payload space is configured around a system of uniformly-sized International Standard Payload Racks (ISPRs). These racks provide1.6 cubic meters of internal volume and weigh 104 kilograms. Destiny has provisions for 23 racks, 6 along each side and the overhead and 5 along the deck. Of these, 13 are ISPRs and the remainder are special purpose racks.

Destiny contains 5 systems racks that provide for such functions as electrical power, cooling water, air revitalization, and temperature and humidity control. Each rack weighs about 545 kilograms. Inside Destiny are racks, rack standoffs, and vestibule jumpers. The rack standoffs provide a volume for ducting, piping, and wiring running to and from the individual racks and throughout the laboratory. The racks connect to the piping and wiring in the standoffs via outlets and ports located in the standoffs at the base end of each rack location. Vestibule jumpers provide for connecting piping and wiring between Destiny and Unity. A Common Berthing Mechanism joins the two modules with the active side on Unity and the passive on Destiny. Each of the two berthing ports on Destiny contains a hatch with a window. The aft hatch to Unity usually remains open. The forward hatch to the PMA is the route to the docked orbiters. The hatches can be opened or closed from either side and they have a pressure interlock feature. This prevents the hatch from being opened if there is a negative pressure across the hatch.

Destiny provides the ISS with the following: Environmental Control and Life Support System: Thermal Control System: Guidance, Control, and Navigation System; Extravehicular Activity; Extravehicular Robotics; Flight Crew Support; Communication and Tracking System; Electrical Power System; Command and Data Handling System; structures and mechanisms; and payload capability.

The Command and Data Handling System gained 11 Multiplexer-Demultiplexers with the activation of Destiny. These are used to control the U.S. on-orbit segment systems. The Communication and Tracking System activated high-rate S-band to replace an earlier communications system. Hardware for the Ku-Band, UHF, and video subsystems were activated during later missions, however the audio system was online with Destiny's activation.

The Environmental Control and Life Support System maintains a pressurized habitable environment within the ISS by supplying correct amounts of oxygen and nitrogen, controlling temperature and humidity, removal of carbon dioxide and other atmospheric contaminants, and monitoring the atmosphere for the presence of combustion products. The system also collects, processes and stores water removed from the atmosphere. The Atmosphere Control and Supply System contains a pressure control assembly, vent relief assembly, and manual pressure equalization valves and gas lines. The Sample Delivery System lines in Unity and Destiny are connected and lead to the major atmospheric constituent analyzer located in Destiny.

In the grasp of the Shuttle's Remote Manipulator System (RMS) robot arm, the Destiny laboratory is moved fromits stowage position in the cargo bay of the Space Shuttle Atlantis.　(NASA)

The Electrical Power System manages, controls, and distributes electrical power to the U. S. pressurized modules. The power is brought to destiny from the P6 arrays through the Z1/laboratory umbilical. The power is brought to two DC-to-DC Converter Units in Destiny and is distributed to the secondary power distribution assemblies and downstream loads.

With the attachment of Destiny to the ISS the U. S. segment began contributing to the attitude control of the ISS with the Control Moment Gyros. The U. S. segment took control of the ISS with state vector and attitude inputs from the Russian segment. The Russian segment propulsive capability was still needed for joint attitude control during Control Moment Gyro desaturation and reboost. The Motion Control System became integrated between the Russian Zvezda and Destiny's computers.

The addition of Destiny added the capability of high-rate S-Band and internal audio to the U.S. on-orbit segment. The S-Band system provided two-way communications with the ISS and Mission Control Center via the TDRSS for commands and system telemetry, vice and file transfer. The internal audio system allows crewmembers to communicate with other crewmembers aboard the ISS

The Quest Joint Airlock

On STS-104 (ISS Mission 7A), launched on July 12, 2001, the Space Shuttle orbiter Atlantis added the Joint Airlock, named Quest. This addition marked the end of Phase Two of the ISS construction. Following the Quest airlock installation the ISS program was capable of supporting EVA operations from the ISS with no Space Shuttle orbiter present. The airlock also provided additional capabilities in that it vents less air into space than the airlock carried in the Space Shuttle's payload bay and the Quest airlock accommodates both Russian and American space suits. The American spacesuits will not fit through the Russian airlocks. Russian spacesuits, however, are compatible with the Quest airlock.

Quest was designed and built by Boeing at NASA's Marshall Space Flight Center in Huntsville, Alabama. The Quest was shipped to the Kennedy Space Center in September of 2000. It was then leak-tested and multi-layer insulation and debris shields were installed on the exterior

Physically, the airlock is spindle-shaped and consists of two pressurized cylindrical chambers along the same axis. These are the Equipment Lock where necessary equipment is stowed and where crew members change into and out of their space suits; and the Crew Lock, from which the crew member leaves and re-enters the ISS. The Crew Lock is 2.56 meters long and approximately 2 meters in diameter. A cone-shaped transition section 0.6 meters long connects the Crew lock to Equipment Lock. The Equipment Lock mates with the Unity Module on the starboard side of that Node. It is 1.9 meters long and 4 meters in diameter. A second cone-shaped transition section necks down to the interface with Unity. Total length of Quest is 5.49 meters. The aluminum airlock weighs 6,064 kilograms and has a volume of 34 cubic meters.

The airlock for the International Space Station is lifted by crane during manufacturing in the Space Station Building. (NASA)

The Quest airlock in the process of being installed onto the starboard side of Unity Node 1 of the International Space Station. (NASA)

The Quest was attached to the Unity on the morning of July 15, 2001 as astronaut Susan Helms, working from inside the ISS, used the station's Canadarm2 to remove the Quest from the orbiter payload bay. STS-104 also brought four high-pressure gas tanks for attachment on the exterior of the Quest airlock. The tanks included two oxygen and two nitrogen tanks, each of which is 0.9 meters in diameter and 1.9 meters in length. They are covered in multi-layer insulation and have debris shields. They are constructed of wrapped carbon fiber, weigh 545.5 kilograms, and have a volume of .45 cubic meters each. The tanks were positioned at 12, 55, 200, and 330-degree clockwise positions from the top of the Equipment Lock during STS-104's second and third EVA.

With the addition of the Quest to the ISS the station weighed in at 118,000 kilograms – approximately the weight of the Space Shuttle.

THE INTERNATIONAL SPACE STATION: EXPEDITION CREWS

NEVILLE KIDGER

The Expeditions

The International Space Station has the objective to become a world class research facility in earth orbit conducting science for all the 16 member countries involved in the construction and operation of the complex.

Eventually it was envisioned that up to seven crewmembers would live at any time aboard the completed complex conducting scientific research in the many laboratories of the ISS and maintaining the systems of the complex.

In order for the ISS to begin useful scientific research it was essential that the facility have a crew on board at the earliest opportunity. Scientific research would begin in the initial phases of the construction of the massive facility.

Continuous Human occupation of the ISS could only begin after the arrival of the Russian Zvezda module, which provided the main life support functions and served as the living quarters for the crew. However, launch of the Russian-made and owned module was delayed for some 2 years because of Russian funding difficulties.

Once the module had been attached to the Zarya/Unity combination in late July 2000 a shuttle mission - STS-106 - was flown in September and the crew spent five days 9 hours and 21 minutes inside installing vital elements of the life support equipment inside the Russian living module including the Elektron Oxygen generator, a floor-mounted treadmill and the toilet. Other tasks included battery and power converter installations.

An EVA was also conducted to integrate cables between the two Russian modules as well as install a magnetometer and boom on Zvezda.

The mission delivered some 2,993 kilograms of supplies and Atlantis' thrusters were fired four times to boost the station's altitude by 22.5 kilometres.

The STS-92 (3A) mission followed in mid-October 2000 and added the Z1 Integrated Truss Structure which would allow for the mounting of the first set of US made solar arrays onto the Unity node to provide more power and a Pressurised Mating Adapter.

Z1 also contained four Control Moment Gyroscopes for control of the complex's attitude without the use of fuel following the activation of the solar arrays. The Pressurised Mating Adapter-3 to allow shuttle Orbiters to dock with the Unity node and the Ku band communications system for supporting science work and television.

At the end of these missions the complex was ready to receive the first in a series of crews of three astronauts and cosmonauts to man the complex in shifts of 3-4 months at a time.

Expedition 1

On 31 October 2000 the first of the crews to keep the ISS permanently occupied was launched from the Baikonur Cosmodrome in Kazakhstan aboard Soyuz TM-31 on a Soyuz-U carrier rocket at 0752:47.241 (UTC).

The Commander was NASA astronaut William Shepherd. He was joined by Russian cosmonauts Yuri Gidzenko (Soyuz Pilot) and Sergei Krikalev (Flight Engineer). The first expedition crew had been training for four years for the mission.

The crew's task would include making the fledgling ISS habitable, discovering and solving problems related to operations of the systems and helping shuttle crews to add the solar arrays and the US laboratory Destiny.

The Expedition One crew. (from left to right): S. Krikalev, Y. Gidzenko and W. Shepherd.

The Back-up crew for Expedition One. (from left to right): V. Dezhurov, K. Bowersox and M. Tyurin.

The crew would also conduct some early science experiments such as earth observations.

Krikalev became the first person to visit the ISS twice. He was a crewmember of the STS-88 mission which began the construction of the ISS in orbit by joining the Unity Node 1 and the Zarya FGB module in December 1998.

The Expedition 1 crew were initially scheduled to return in February 2001.

The crew received two unmanned Progress cargo spacecraft which carried consumables to the complex and relocated their Soyuz TM ferry craft from the Zvezda aft docking unit to the Zarya nadir port.

They also received two assembly Space Shuttle missions:

- STS-97 in December, with the first set of the huge solar arrays to provide power for the complex;
- STS-98, in February, which delivered the vital Destiny laboratory.

The STS-98 mission was delayed by about 3 weeks to allow further inspections of cables on the Solid Rocket Boosters of the shuttle stack, extending the planned stay of the first expedition.

In the event the Expedition 1 crew's return to earth was delayed until 21 March 2001 because of the delay in launching STS-98/5A.

Early Operations on ISS

Soyuz TM-31 carrying the Expedition 1 crew was docked with the rear port of the Zvezda module on 2 November and Shepherd, Gidzenko and Krikalev were soon inside the Russian module.

In a TV report, just after entering Zvezda, Shepherd asked for permission to use the call-sign "Alpha" for the rest of the Expedition 1's mission. NASA Administrator Daniel Goldin in the Korolev TsUP gave permission to the experienced NASA astronaut.

Despite the fact that no official name had been agreed for the ISS by the international partners the "Alpha" call-sign became the permanent one for the following expeditions. By the end of the third increment the complex was still without an official name.

In the first days aboard the complex the Expedition 1 crew connected and started the Vozdukh carbon dioxide removal system, the Elektron oxygen generator and the SKV-1 air conditioning system (one of two).

The crew used lithium hydroxide canisters to remove CO2 and burned modified solid canisters to produce oxygen during the early days until the systems were operational about a week into the crew's mission.

Shepherd kept a log of the crew's activities and NASA published a series of censored versions on their public web site after the mission. The logs detailed the daily frustrations of the crew in establishing the complex as an operational base.

Shepherd noted that the crew were, after all, the pioneers of the ISS.

Working to a Routine

The crew worked to a timetable based around Greenwich Mean Time (GMT). Their days started at 0600 GMT and officially ended at 2130 GMT although many days lasted longer because of problems or the volume and complexity of the work that was required.

The operational problems that were encountered during the early phase of the crew's stay, during which they worked the long hours, caused the crew to complain that the schedulers on the ground had crammed too much work into their days.

The crew instigated a "job jar" principle for tasks that needed to be performed but could not be timelined on the official daily Russian Radiogram or the US plans which were sent to the crew through the communications links each day. The tasks could be performed at the crew's discretion and often this involved using part of their leisure time up for these tasks.

Daily routine tasks for all the crews included maintenance of the SOSh, or Russian life support systems, such as the Elektron oxygen generator, attaching water containers for the system to produce oxygen, inspection of the water condensate separator, air conditioning system and the Vozdukh carbon dioxide removal system.

The crew produced computer files of the results of radiation detectors and their medical data and returned them to the Earth via telemetry link on a regular basis.

Personal hygiene (wipe down with damp cloths and wet shaving) took place in an area of the crowded Zarya module.

Other routines included a clean-up of the grilles of the air circulation and filtering systems on a weekly basis and a weekly wipe down of open surfaces with wet napkins to clean off microbial growth.

The ASU toilet at the rear of the Zvezda module was also serviced on a regular basis.

Filters and fans required changing often, to a schedule, and the Expedition 1 crew had to work with errant battery blocks inside the Zvezda and Zarya modules.

The men returned electronic still pictures of their activities both for the experts on the ground and for investigators and the general public.

All of the ISS crews were scheduled for weekends as rest days and holidays, such as Christmas and Easter were observed with rest days. Often, on passes over the Russian ground stations, the Moscow Control centre (TsUP) would broadcast Russian news reports to the crew to keep them informed.

On weekends personal conferences were organised with family members and friends.

Expedition One Commander William M. Shepherd floats in the Zvezda Service Module.   (NASA)

Problems Found, Problems Solved

The logs of William Shepherd, even in their censored form, provide a fascinating glimpse into the trials that the three cosmonauts were presented with in order to configure the station (especially the Russian Zvezda module) into a "home". The logs also detailed the repair work that sometimes had to be accomplished outside of the normal setups and checkouts of the new module.

Problem solving of all kinds was hampered by the intermittent communications with the control centre in Korolev, near Moscow, restricted to the periods when the station was in line of sight of ground stations.

After one particularly frustrating day early in the mission Shepherd wrote:

> "In summary, productivity of the day was at best, modest. Limited availability of comm is having a significant impact on the crew's ability to work through minor issues with the ground team's help. We need a little advice here and there to keep the flow of work moving, but we just have not been able to talk enough. We are probably keeping the planners busy as they reschedule what we didn't finish. We will try to make things work as smoothly as we can."

Zvezda's air conditioning unit (SKV, one of two loops) was found to have blobs of water accumulating and at one stage during Expedition 1 a large blob escaped from a valve. The crew used napkins to clean up the liquid.

Shepherd later wrote:

> "…regarding the condensate on the exterior lines for SKV. Unit is producing fairly sizeable bubbles of water-maybe 50g each. We had a similar problem on SKV-2 which was readily fixed with some insulated wrapping and a bit of airconditioner tape, and we are ready to do this again."

On 1 December Shepherd wrote, about the exchange of air conditioning components following the disintegration of a fan in the SKV-2 loop:

> "Fit of heat exchanger unit is very tight. We are saying bad things about the engineers. Sergei is sure the unit has been fit-checked on the ground. You can see bright marks where parts were binding together, though. Files, silicon lube, hammers, pry bar, line up tool all in the fight. Finally have to loosen the support bolts for the whole airconditioning unit to float it a little bit and get enough space to slide the heat exchanger in. Could have used the "large vise grips" at least twice, and a 1/4" die-grinder would have been real handy. Told (the ground) that so they would sense this was not your ordinary "bolt it down" job. More like changing out your transmission."

Later that week, following the departure of the STS-97 crew, who had delivered a pair of vice grips to the veteran astronaut, Shepherd wrote:

> "… starting to put notches in the large vise grips. Used them again today to loosen the vacuum jumper line from the MPEV valve. Even though the hose says hand tighten only-doesn't mean you won't need a tool to get it off. Just think it should be SOP on station that we have tools for every fitting we are expected to work with."

At the end of December Shepherd described another activity which took longer than planned:

> "The workday starts early as usual. Sergei getting ready to install the GPS/GLONASS satellite receiver (ACH). He has about 1 ½ hours scheduled to do this, but the installation is way up on panel SM 338 (i.e. the "ceiling" of Zvezda). Access is not good, and lots of cables look like they were routed after the ACH box was installed. They are all in the way. Sergei takes an hour just to scope out the installation and plan how to do it. Job takes 4+ hours, but Sergei comments that it would have been a full day's work if he had just started it without making a good plan."

Shepherd wrote in mid-January after a long day of inflight maintenance work on connectors for a communications head-set:

> "In all, we have recently experienced-a scopemeter we can't recharge because we can't plug it in, a soldering iron which has the wrong size tips, which we can't plug in either, a vise we can't use because it's still on the ground, and "rivnuts" we can't use because we don't have the right drill bits. Not to mention a workbench that's still on the ground somewhere too. We are enjoying finding all these "surprises" particularly before we would need these tools to do something critical. It would be nice, however, just to be able to pull these things out and start using them. We would like to encourage a better process to integrate and check this gear before we see it."

Computer Setup Problems

The Expedition 1 crew had to route wires and locate the laptops for the Station Support Computer (SSC) network in the Zvezda module and test their functioning with the central computer complex of the module.

This meant the crew had to work with a European central computer of the Zvezda module, US and Russian made laptop systems.

Computer systems oversaw the functioning of the systems of the complex and their reliability was a major key to the success of the ISS project.

For the first weeks much of the first crew's work days were spent troubleshooting the computer software and many of the files which were transmitted to the complex through the early communications channels were lost in the computers.

Sergei Krikalev (left) and Yuri Gidzenko (right) at work in the Zvezda Service Module.   **(NASA)**

Sergei Krikalev combines work and eating aboard the International Space Station. (NASA)

Even well into the mission the crew were reporting daily problems in receiving and sending e-mails.

According to reports from an observer familiar with the censored parts of the Shepherd logs the crew complained that a computerised system called a "Crew Squawk" which was used to record problems and report them to the ground was itself experiencing problems communicating.

Making Zvezda into a Home

Shepherd and his crew remarked upon the amount of times they used their power tools for conducting repairs and to remove covers and access panels.

Metal cutters were also used extensively as Shepherd fashioned a "table" for the Zvezda module out of frames and containers holding items in the Progress cargo ship until the ground could find the room to fly the proper (larger) table up to the complex in a later Progress.

The crew would meet at the table to eat meals and discuss plans and day's events. In the evenings the men would watch films on CD-Rom and DVDs on a portable computer.

Shepherd wrote, after the delay of the STS-98 mission from mid-January to early February, that he was worried that the supply of good films to watch would run out. He joked that the trio would be able to turn off the sound of the films and "mouth" the dialogue because they knew some films so well. Shepherd commented that it was strange to watch a film about space travel (2010) whilst actually flying on one.

Unity off Limits

The crew were initially unable to use the full volume of the fledgling ISS complex because the Unity node was sealed to allow the power that would heat that module to be diverted to keeping the four huge CMG gyroscopes on the Z1 truss segment heated before the power producing Solar arrays arrived in December on the P6 structure.

Once the huge arrays were installed and producing power the crew were able to enter the node and greet the crew of the STS-97 mission who had brought the solar arrays.

With access to the large volume of Unity the crew could place extra equipment in that module and remove items that had been stowed there by earlier shuttle crews. One of the vestibules of a docking port of the node was used to store the large volume of water bags that the shuttle crews had delivered.

To illustrate the continuing small problems that the crew encountered, an extract from Shepherd's 5 February log entry reveals:

> "Working the TCS QD's(Thermal Control System Quick Disconnect fittings); none of the fittings have the thermal mittens on them as shown in the pictures. Everything disconnected, bagged and taped according to the new procedure. The annotations are good and very clear. Our thanks to the folks who jumped on this. The Node panels are still very difficult to reinstall and fit up right-even with a couple of fasteners threaded up, many of the remaining ones fail to "find" their captive nuts. We noted that on the P1 panel, there is a lengthy 3 page fit up procedure "decal" which explains how to put the panel back on. Unfortunately, it's on the back side of the panel."

For two weeks the crew kept bags of seeds in Unity as part of student experiment.

Exercises to Keep Fit

The Expedition 1 trio did not perform any of the recommended physical exercises during the early part of their mission because of the high humidity in Zvezda.

The routine that the crew eventually followed saw them perform exercises for up to 3 hours per day on the TVIS vibration isolated treadmill installed in the Zvezda module (between the two sleep compartments), an Interim Resistive Exercise Device (essentially a weightlifting frame which was eventually located in the Unity node) and the Zvezda stationary bicycle, the veloergometer.

There were early problems with the operation of the veloergometer and with static build-ups on the TVIS control panel imparting electric shocks to the subject exercising.

After engineers on the ground had recommended using straps for grounding the charge, Shepherd wrote on 31 January:

> "We are using the wrist bracelet rigged backwards as a means to provide grounding. The alligator clip is attached to clothing, and the wrist strap is on the bottom of the TVIS control panel. This prevents the grounding wire from interfering with running motion on the treadmill. As an interim solution for the static charge build up, this is working."

The crew measured their body masses and measured their calf volumes on a regular basis for medical evaluation. Shepherd noted in early January that the measured values corresponded with their pre-flight ones and gave the credit to the amount of food the crew were consuming.

Blood samples were also given by the crew for analysis on board using a small centrifuge and a portable blood analyser.

In December the Russian cosmonauts began using the Chibis pressure leggings to simulate Earth's gravity for medical tests.

Keeping Track of Things

Following the access into Unity, the need to keep track of items which were used on a daily basis and also items for future expeditions, which needed to be stowed and later found, became a major issue.

Many hours were spent wrestling with a computerised bar code based inventory management system (IMS). Some times items which were sent to the computers from the ground were updated with the wrong data, leading to items not being found.

The shuttle missions which visited the ISS before the arrival of the Expedition 1 crew had delivered about 10,000 items which had been stored in the various modules, especially the FGB and the Node.

In reports similar to those from previous Russian and American long duration missions Shepherd noted that items which had been put aside had floated away from the crewmember and often found their way into or on the filters of the modules, carried there by the flow of air through the complex:

> "Went thru the ship with the vacuum cleaner-pulled all the debris out of filters and intakes. It is amazing-if you ever let

loose of something, there is an almost 100% chance it is going to end up in a filter or screen somewhere. So far, we have not "lost" anything important, except maybe some "1"s and "0"s in the IMS database. "

Acoustic Environment of the Russian Segment

Excessive noise in the Russian segment in particular, from fans and operational equipment, was recognised as a problem before the launch of the Zvezda module.

Shepherd's crew spent much time working with acoustic meters to categorise the "noise hotspots" of the complex.

Even though some muffles had been placed around some equipment which had been expected to be noisy the findings were that the complex was still a noisy environment in which to live and work.

Shepherd wrote:

> "Logging acoustic data on MEC. Numbers are roughly 61-63 dB around our sleep locations, 75 in work areas and central post, and 80-85 around the noisiest equipment."

In later debriefings the crew reported that noise would be an on going problem for the crews of the ISS. For example the Vodzukh CO2 removal system was intermittently very noisy in Zvezda and it was also located close to the sleeping rooms of two of the crew.

Progress Craft

The first unmanned cargo craft to be launched to the Expedition 1 crew was Progress M1-4.

On 18 November the docking of the unmanned craft to the Zarya FGB nadir docking port allowed Soyuz Pilot Yuri Gidzenko the opportunity to show off his remote flying skills when the ship's automated Kurs rendezvous system failed.

Inside Zvezda was the TORU unit for controlling the approach of an unmanned cargo ship using a joystick control panel and TV beamed from the approaching craft. Gidzenko successfully docked Progress after the Kurs system experienced problems during a switchover from the Zvezda to the Zarya antennas.

The Progress delivered more food, clothes, fuel for the Zvezda and Zarya tanks and oxygen which the crew released into the complex's volume at intervals to maintain the correct pressure.

Progress M1-4 was undocked just after the launch of STS-97 and kept in orbit so that Gidzenko could practice the TORU controlled docking after the departure of the shuttle Orbiter to check out the software fixes made by the controllers at TsUP.

That docking was successfully accomplished on the morning of 26 December.

Progress was undocked for the second and final time on 8 February and sent to a destructive entry into the Earth's atmosphere loaded up with used equipment and rubbish.

Shepherd wrote:

> "The cargo ship moves away below us-it's quick compared to a Shuttle departure. We can see the outline of the spacecraft—the spotlight is pretty bright. Even though it is a night pass over Russia, the snow and cloud cover make the Earth background very visible. It looks a lot like twilight although we are well past the terminator. The engine pulse on Progress is also very visible and vehicle starts a slow rotation as well. We're observing thru the large deck window. We try to get some video images on the camcorder but the light level is very low. The Progress moves below and a little bit forward on us. It is just a black speck now against the cloud cover. The only way we can pick it up is that it does not appear to move with the cloud cover. Sergei is using the laser rangefinder to get some trajectory data on the departure."

The second cargo craft, Progress M-44 was docked to the rear Zvezda port late in the mission of Expedition 1 after the crew had relocated their Soyuz TM-31 from that port to the vacated Zarya nadir port.

EVA

An originally planned internal EVA to move the Zvezda docking cone unit from the front port, where it was used to connect with Zarya to the nadir port of the module for the docking of the Russian Docking Compartment-1was planned for December 2000 but was deferred to Expedition 2.

In Early January the Russian cosmonauts checked out the Orlan EVA suits and found a communications problem with them which needed further analysis. Again, Shepherd noted that the amount of time set aside by the planners for the tasks involved had been underestimated,

Ham Radio

In contrast to the often poor communications with the ground through the early communications systems channels the crew complemented the operation of their onboard amateur radio system.

Located in the Zarya module the crew were able to use the rig to speak with radio amateurs, schools and family members.

Traditions

The Expedition 1 crew established traditions, some base upon naval ones, for the resident crews to observe during visits from other spacecraft.

These included the ringing of a "ship's bell", brought by the STS-106 crew, to announce the arrivals and departures of other spacecraft and the placing of mission emblems on a portion of the wall of the Unity Node 1.

The ringing of the ship's bell was usually done by a US crewmember.

Destiny in Space

The final 6 weeks of the Expedition 1 crew's stay in space was busy with the arrival of the Destiny science laboratory and their replacements, the Expedition 2 crew, on two separate shuttle missions.

Atlantis docked with the ISS on 9 February 2001 to delivery the Destiny laboratory which was attached to the forward port of the Unity node the next day.

11 February was a banner day for the ISS programme. After signing an official form taking ownership of the new piece of the ISS the two crews, led by Shepherd and Atlantis commander Kenneth Cockrell led their crews into the spacious volume of the laboratory.

Shepherd wrote:

"The lab is well lit, very spacious, and clean. (probably the last time we'll be able to say that). We spend several minutes just looking around our new home. Even with many racks yet to be plugged in, the lab is a very impressive piece of hardware."

Describing the initial work inside the laboratory he continued:

"Everyone is climbing over racks and closeouts. Marsha is keeping all the action pointed in the right direction-Shep and Yuri chasing tools and parts and passing bags, Taco pulling out plumbing, Sergei, Beamer, and Roman working in the endcone on ducts and valves. Tom on the videocam. All is very much like the SEAL saying— "two shooting, two looting, and one taking pictures.""

(Shepherd was a US Navy SEAL before becoming an astronaut.)

The crews activated air supply systems, fire extinguishers, alarm systems, computers and internal communications systems during the early hours of the Destiny activation.

By day's end the crews had completed a large part of the complex lab setup activities. Shepherd's log recorded:

> "We can't say enough about the good work of the Atlantis crew—They came, they saw, they installed. All in all, it was a landmark day for the station. We now have what looks to be at least 40 meters of open hatchway from the SM wardroom to the front of the lab."

The two crews made great use of the large-format IMAX 3-D camera during the joint portion of the flight photographing sequences of work inside, and outside the complex which would be later used in a spectacular film of the ISS construction.

On 13 February the attitude control of the station was passed to the large CMG gyros on the Z1 truss.

Transfer of items between the shuttle and station was completed on 15 February. In all some 1362 kg of equipment and supplies – water, food, spare parts, a spare Vozdukh carbon dioxide removal system, computer, clothes, movies and other items – were moved from Atlantis to the Station. About 386 kg of rubbish and other items were moved from the ISS to Atlantis.

The crew thanked the ground for sending up new movies and a DVD player and some new software for the Russian computer system including a new "World Map" programme to show where the station was above the planet.

A second software application gained Shepherd's approval, saying it was:

> "outstanding even by Russian standards-a combined bird's eye view, world map, and star field which runs on the Russian PCS. It also reads state vector and attitude quaternion updates directly on the bus. Very slick. (Now if we could get a couple of large format LCD screens to hang in the front of the Central Post, we could give "Enterprise" a run)."

STS-98 left the station on 16 February leaving Shepherd, Gidzenko and Krikalev to complete their work before the next crew arrived on the station.

Final Month Begins

A maintenance procedure to replace some broken slats on the TVIS treadmill was part of the first order of business for the crew after the excitement of the Destiny arrival. After a long struggle the slats were bolted down, only for another one to fail the same day during the next exercise session. On 25 February the crew took the TVIS out of the pit in which it was located in the Zvezda floor to make a full repair of the slat problem.

The crew also continued the configuration of the computer networks complete with problems in which the connections between the computers would not work for a period of several hours, described at length by Shepherd.

The crew also continued taking yet more sound meter readings with the comment that these continued operations were:

> "way past our next most fun thing to do now."

Redocking Soyuz

24 February saw the re-docking of the Soyuz TM-31 spacecraft from the Zvezda rear port to the Zarya nadir port.

Future Progress craft would now dock at the Zvezda rear port and Soyuz craft at the Zarya nadir port. After the arrival of the Pirs docking compartment and its attachment to the Zvezda nadir port Soyuz craft would be moored at those two ports during ferry ship swap flights.

Shepherd's log described the re-docking exercise. After closing down systems of the Zvezda and Zarya the trio donned their Sokol suits and entered Soyuz for the undocking preparations. The American commander wrote that he had forgotten how small the Soyuz seats were. During the waiting (the re-docking had to take place over the coverage area of the Russian ground stations) the crew watched a DVD film.

> "We finally are ready to do the second set of hooks and undock just after 1000. Comm is through Sband link now and we aren't getting voice up from Moscow. Yuri broadcasts in the blind anyway. We come off the docking adapter pretty

briskly, and we are moving straight aft. It's kind of surprising how little rotation we have picked up in the separation. We drift out to 30 meters or so and Yuri starts flying manually. We have a timeline where (Moscow) wants us to be based on comm. coverage and lighting, but we are not real happy that there seems to be no real loss of comm. plan once we are off. We had asked about this, but were told that we should just "sit tight" till we come back into radio contact. We pick up Moscow the radio. Yuri does his usual great job flying us around and we make a short trip up the underside of the station. Visibility, lighting, target contrast are all OK. Then we get to hang out for the final approach. Moscow gives the "go", and we are moving in. About 3 meters out, we get a "wait" call as the TV picture on the ground is suddenly bad. Nobody's very happy about this-it's a bad place for a "wait-out". Fortunately, Moscow comes back quickly and gives us a go.

"We make contact-it's a moderate bump with the docking mechanism. Then the docking program starts working to tighten everything up. Then a second set of hooks from FGB, and pressure checks. More DVD. We have a good seal, and we equalize the docking tunnel, and then to the station."

On 28 February the Progress M-44 craft docked with the rear Zvezda port.

In the final days of February and early March the crew set-up and began operations with the Plasma-Crystal experiment.

Flight Events of Expedition 1 at a Glance

Duration (launch to landing): 140 days 23 hr 38 min 55sec

Date	Event
October 2000	
31/10	launch of Soyuz TM-31 from Baikonur
November 2000	
1/11	Progress M1-3 undocked from Zvezda and de-orbited
2/11	Soyuz TM-31 docked with Zvezda (automatic)
16/11	launch of Progress M1-4 from Baikonur
18/11	Docking of PM1-4 to Zarya nadir (using TORU)
December 2000	
1/12	STS-97(Endeavour) launched with P6 solar arrays
1/12	Progress M1-4 undocked and placed in parking orbit
2/12	Endeavour docked to PMA-3 port of ISS
9/12	Endeavour undocked
26/12	Progress M1-4 re-docked (using TORU)
February 2001	
7/2/01	STS-98 (Atlantis) launches with the Destiny laboratory
8/2/01	Progress M1-4 undocked and de-orbited
9/2/01	Atlantis docked with ISS at PMA-3
16/2/01	Atlantis undocked
24/2/01	Soyuz TM-31 undocked from Zvezda rear port and re-docked to Zarya nadir port
26/2/01	Progress M-44 launched
28/2/01	Progress M-44 docked at rear Zvezda port
March 2001	
8/3	STS-102 (Discovery) launched with Expedition 2 crew
10/3	Discovery docks with ISS PMA-2 port
19/3	Discovery undocks with Expedition 1 crew
21/3	Discovery lands Expedition 1 ends

Expedition 2 Arrives

STS-102 carrying Expedition 2 was launched on 8 March and docked with the complex two days later.

The shuttle also brought another part of the ISS infrastructure to orbit – an Italian-made logistics supply module (essentially a cargo container) which was docked with the Unity node after an EVA by Expedition 2 crewmembers Voss and Helms.

The large container, called Leonardo by the Italians, contained the first science rack for Destiny and control

The Expedition Two crew. (from left to right): S. Helms, Y. Usachev and J. Voss.

The Back-up Expedition Two crew. (from left to right): C. Walz, Y. Onufriyenko and D. Bursch.

stations for the imminent arrival of the Canadian robotic arm, Canadarm2, which would be essential for the further construction of the complex.

Exedition 2

The Expedition 2 crew consisted of veteran Mir Flight Engineer, now Commander, Yuri Usachev, with two experienced American astronauts James Voss and Susan Helms.

The Expedition 2 crew had already visited the two-room fledgling ISS on the STS-101 mission in May 2000 and worked in the Zarya FGB and the Unity Node.

The change over from Expedition 1 to Expedition 2 took 3 days during the shuttle mission as each of the participants assembled their Individual Equipment Liner Kit (specially moulded seat) into the Soyuz TM-31 lifeboat.

It was the only time that this staggered change of personnel was used in the first year of ISS occupation.

The work of the second expedition began with a false fire alarm from the Destiny laboratory shortly after the departure of STS-102 on 19 March. The false alarm was troublesome for the crew because they were initially unable to locate the written procures on how to deal with an onboard fire. Computer versions of the procedures were available on the complex's network.

Usachev and Voss moved into the two sleeping rooms in Zvezda and Helms found an empty rack space inside Destiny which became her sleep station.

The crew also installed another exercise bicycle, the CEVIS, inside the Destiny laboratory.

Other early tasks for the newly arrived crew involved unloading of the Progress M-44 cargo spacecraft.

Science Work Begins

The second crew to work on the complex began the programme of science work in earnest by installing some of the first commercial processing experiments in the racks of the Destiny laboratory and installing a suite of three radiation detectors in the laboratory during late March.

One of these experiments was the so-called "Phantom Torso", a mock-up of a human torso, although it contained no human tissues, which was placed in an empty rack space in Destiny.

In total the crew were to conduct 18 experiments as a part of the American programme and continue work on over 20 Russian projects, most of which were untended experiments but also included Earth observations and commercial activities.

Helms spent many of her spare hours working with the MACE-II experiment inside the Unity node. The experiment tested the control mechanisms for evaluation in large articulated spacecraft and other industrial applications.

Other experiments included initiation of a long series of tests with the ARIS-ICE vibration isolator rack, medical tests on the human body and questionnaires which used the Johnson Space Centre's Human Research Facility Rack (HRF).

Susan J. Helms pauses from her work to pose for a photograph.   (NASA)

Amongst the other improvements for the Expedition 2 crew were the much improved Communications links, using the Ku band system which allowed high quality TV to be returned After early problems with the system, which were resolved by new software sent to the complex through radio channels, the system was activated in mid-April.

Progress Departs, Soyuz Redocking

Expedition 2 saw the arrival of one Progress cargo ship and the re-docking of the Soyuz TM-31 craft which had brought Expedition 1 to the Zvezda nadir port, leaving the Zarya port free to receive a new craft, Soyuz TM-32.

ProgressM-44 was undocked and destroyed in the atmosphere on 16 April

Soyuz TM-32's redocking took place on 18 April 2001 with the ferry ship being moved from the Zarya FGB nadir port to Zvezda's rear port.

Yuri Usachev manually flew the Soyuz craft to the redocking without incident. Before such operations the crew had to turn off much of the equipment of the complex in case they could not perform the docking. Once back inside the station was reactivated. The whole operation took the whole crew the best part of a day to complete.

Soyuz vehicle swaps were a regular occurrence during the Salyut and Mir missions due to the approximately 6 months of certified operational lifetime of the vehicle in space.

Soyuz fulfilled the function of a crew lifeboat in the absence of a Space Shuttle Orbiter. The presence of just one Soyuz meant that the Expedition crew size of the complex was restricted to 3 people because that was the maximum that one Soyuz could carry.

Robotics – A Major Focus of Expedition 2

The Expedition 2 crew extensively tested the operation of the Canadarm2 Space Station's Robotic Manipulator System (SSRM) using a pair of robotics work stations (RWS) in the Destiny laboratory. The RWS had been delivered and installed during the STS-102 mission and the crew tested them out before the arrival of the SSRMS on the STS-100 mission in April.

James S. Voss and Susan J. Helms work at the controls of the new Canadian-built space station robotic arm.   (NASA)

Over the weeks following the installation of the SSRMS on Destiny, and the resolution of the command and control computer failures (see below), Voss and Helms conducted a series of weekly tests of the functioning of the arm to prepare for the first major task the arm would perform the movement and attachment of the massive joint airlock to a port on the Unity node.

During these checkouts the astronauts discovered a major with the operation of the arm in the primary control mode (the arm had two control "strings" - prime and reserve).

On one occasion, after the arm had been attached for a time to one of the Power and Data Grapple Fixtures (PDGF) on the Destiny laboratory, the release mechanism stalled and the arm contacted the laboratory and caused a "bang" which momentarily startled the crew inside.

The problems and the repeated series of tests of the functioning of the SSRMS (including operation on the reserve string only for the attachment of the airlock) lead to weeks of uncertainty about the certification of the system and a corresponding delay to the launch of the STS-104 mission which would deliver the airlock.

A later mission would see a suspect wrist roll joint exchanged on the Canadarm2 system.

Computer Failures

During the visit of the STS-100 crew, delivering the SSRMS, the three controlling computers in the Destiny laboratory, known and command and control computers (C&C 1, 2 and 3), all experienced hard drive failures which took them offline from controlling the activities of the US Segment.

Fortunately the failures took placed whilst the Orbiter was docked and the crew was able to work with the vast ground support team of the American agency to reboot the trio of control computers before the departure of the visiting crew of astronauts and cosmonauts. The problems meant an extended stay at the ISS by the STS-100 crew before they could complete their mission, leaving the Canadarm2 attached to Destiny, and undock.

The computer failures resulted in the replacement of the hard drives of all three of the computers by solid-state units flown later in the year.

Taxi Mission 1

Late April and early May 2001 saw the controversial visit to the ISS, and the Expedition 2 crew, by the American businessman Dennis Tito.

Tito rode on the Soyuz TM-32 spacecraft on a mission to swap Soyuz ferry ships with the one which flew Expedition 1 to the complex.

The Commander of Soyuz TM-32 was Talgat Musabayev with Flight Engineer Yuri Baturin.

NASA's then-administrator forbade any TV coverage by his organisation of the onboard activities of the businessman who had paid the Russian side a reported 13 million dollars for his trip to the station. Russian TV systems in the Zvezda module were used for the coverage of the arrival of the crew and for TV reports and interviews during the short stay on the complex by Tito and his colleagues (It should be noted that NASA said there were problems with the Ku band system and the more limited OCA system in any event at this time.)

Originally Tito had sought to fly to the Russian Mir complex but the demise of that station in March 2001 left the Russians ready to honour their commercial agreement, despite the misgivings of the rest of the international partnership, by flying him to the ISS instead.

The crew was launched on Soyuz TM-32 on 28 April 2001, while the Orbiter Discovery was still docked to the ISS. The Russians had refused to delay the launch of the Soyuz because of the extended period the Orbiter was docked to the complex (due to the three computer failures).

Soyuz TM-31 was undocked on 6 May and returned to Earth leaving the Expedition 2 crew to continue their mission.

Progress M1-6

The next cargo ship, Progress M1-6 was launched from Baikonur on 20 May and docked with the recently vacated rear port on Zvezda on 23 May delivering 2478 kg of cargoes to the complex.

As the crew began their unloading sequence of the Progress craft NASA announced that the duration of the crew's stay was to be extended as engineers began the process of studying the problems being experienced with the Canadarm2.

EVAs

James Voss and Susan Helms had conducted an 8 hour 56 minute EVA from the shuttle Discovery during the hand-over phase of the STS-102 mission. (The pair's tasks included unplugging Pressurized Mating Adapter 3 from the Unity Node electrical cables, Lab Cradle Assembly installation and rigid umbilical installation. The 17th space walk in the space station assembly sequence was also the longest in Space Shuttle history.) That EVA is covered in the Shuttle Assembly portion of this book.

The other EVA of Expedition 2's ISS occupation was the relocation of the Zvezda docking cone from a storage site in the module's docking node to a position on the nadir docking unit.

The EVA (although neither participant ventured outside the depressurised Zvezda node) was conducted in Russian Orlan suits and was performed by Yuri Usachev and James Voss on 8 June 2001. It was designated EVA-1 by the ISS programme, for the first Russian Segment EVA.

View of a Russian Orlan space suit drifting in the Zarya module.   (NASA)

Sealed in the Zvezda docking node the pair removed a plate from the nadir docking port and attached the cone which had originally been on the front docking unit of Zvezda.

Once the cone was attached the pair re-pressurised the node and moved back into the modules. Helms remained inside the US segment, in case she needed to evacuate to the Soyuz TM-32 escape craft docked to the Zarya nadir port.

The total length of the EVA was just 19 minutes.

Quest Arrives

The highlight of July for the Expedition 2 crew was the arrival of the joint airlock, christened "Quest" at launch on the Orbiter Atlantis on Mission STS-104.

The Orbiter Endeavour docked with the front port of the complex on 14 July and the next working day the weeks of training with the Canadarm2 paid off.

Using the Canadarm2, controlled by Voss and Helms, the airlock was extracted from the Orbiter's payload bay and positioned on the Unity node where it was bolted to the structure (aided by two shuttle EVA crewmembers) to become a permanent part of the complex.

Endeavour undocked on 21 July after a successful EVA by STS-104 astronauts Gernhardt and Reilly from the Quest airlock the day before to inaugurate the operation of the facility.

End of Expedition 2

As the date for their delayed return to Earth approached the Expedition 2 crew continued their daily cycle of science activities, medical tests and systems maintenance.

Russian segment activity involved the raising of the orbit of the complex with the engines of the Progress craft and the replacement and stowage of the Kurs automated rendezvous systems of the Progress and Soyuz crafts which were to be returned to Earth for reuse on later vehicles.

Flight Events of Expedition 2 at a Glance

Duration (launch to landing): 167 days 6 hr 40 min 49 sec

Date	Event
March 2001	
8/3	launch of STS-102
10/3	docking with ISS
11/3	EVA to reposition PMA-3 (Voss, Helms) 8:56
19/3	undocking of Discovery
April 2001	
16/4	undock of Progress M-44, descent into atmosphere
18/4	re-docking of Soyuz TM-31 from Zarya FGB nadir rear to Zvezda Aft port
19/4	launch of STS-100 (Canadarm2)
21/4	docking of Endeavour
28/ 4	launch of Taxi-1 on Soyuz TM-32
29/4	undocking of Endeavour
30/4	docking of Soyuz TM-32 the Zarya FGB nadir port
May 2001	
06/5	undock and landing of Soyuz TM-31 with Taxi-1 crew
20/5	launch of Progress M1-6
23/5	docking of Progress M1-6
June 2001	
8/6	EVA-1 (Usachev,Voss) 19 minutes
July 2001	
12/7	launch of Atlantis STS-104 (Quest)
14/7	docking of STS-104
22/7	undocking of STS-104
August 2001	
10/8	launch of STS-105 (Expedition 3)
12/8	docking of Discovery STS-105
20/8	undocking STS-105
22/8	landing of STS-105 with Expedition 2 crew

Expedition 3 Arrives

STS-105 was launched on 10 August and docked with the ISS at the front PMA port two days later bringing the third expedition crew and several tonnes of supplies in the Leonardo MPLM.

The day after docking the seatliners of the Soyuz ship had been exchanged and Expedition 2's tenure at the complex was officially at an end. The third expedition began their stay which was scheduled to last until early December.

The Expedition Three crew. (from left to right):
F. Culbertson, V. Dezhurov and M. Tyurin.

The Back-up crew for Expedition Three. (from left to right): V. Korzun, P. Whitson and S. Treshchev.

A highlight of the hand-over period between the two expeditions was the uploading of new software for the control computers of Zvezda.

Expedition 3

The Expedition 3 crew comprised American commander Frank Culbertson, past US manager of the Shuttle/Mir programme, and Russian cosmonauts Vladimir Dezhurov and Mikhail Tyurin.

The two Russians, together with American astronaut Kenneth Bowersox had been the reserves for the Expedition 1 crew. Culbertson had replaced Bowersox on the Expedition 3 crew after that assignment had been fulfilled. Bowersox was later to command the sixth Expedition to the ISS.

The Expedition 3 crew did not host any visiting Space Shuttle missions for further ISS assembly during their stay but they did oversee the next major addition to the complex and received the second visiting Taxi mission which featured the flight of a French researcher.

V. Dezhurov works on a laptop computer in the Temporary Sleep Station (TSS) in the U.S. Laboratory Destiny.   (NASA)

Science Programme

The science programme of the Expedition 3 crew incorporated about 20 US experiments including new elements as the programme expanded. The study of bone mass and muscle atrophy was a major part of the Expedition 3 science focus.

The Russian science programme saw the continuation of the untended equipment recording radiation and micrometeoroids around the complex.

Two more EXPRESS Racks were delivered with Expedition 3 and installed in Destiny.

First Progress Ship

The first Progress craft launched to Expedition 3 was Progress M-45 on 21 August. The spacecraft docked with the rear port of the complex on 23 August. It delivered some 2.5 tonnes of equipment to the complex including more than 890 kg of propellant for refuelling of the complex, 210kg of water and 1420 kg of dry cargoes including experimental materials for the Japanese and French science programmes.

The Progress M1-6 craft had been undocked from the complex a day after the launch of M-45 and was deorbited over the Pacific Ocean shortly afterwards.

September 11 2001

The aftermath of the terrorist atrocities in New York and Washington D.C. were witnessed by the crew of Expedition 3 who were informed of the events during a medical check-up late in their working day. Shortly afterwards the complex flew over the area of New York and the crew were able to see the huge pall of smoke and dust from the destroyed World Trade Centre buildings.

Frank Culbertson wrote in a touching "Letter Home", published later by NASA, that he had been a classmate of the pilot whose hijacked aeroplane had impacted the Pentagon in Washington D.C.

The crew were given regular daily updates from 11 September onwards of the ongoing operations at the sites of the outrages and the new "War on Terror".

Pirs Docking Compartment

The Docking Compartment-1 (Called Pirs – "Pier" - by the Russians) was launched towards the complex on 14 September from Baikonur on a Soyuz-U carrier rocket. The module was attached to the front of a modified Progress craft and was named Progress M-SO1.

Progress M-SO1 docked automatically and without incident to the Zvezda nadir port (at the docking unit prepared by Usachev and Voss of Expedition 2 on 8 June). The unneeded Progress engine unit (designated the instrumentation and propulsion section) was discarded some days later.

The Pirs module was to be used by crews for EVAs from the Russian segment and was outfitted by the crew for first use by that crew the following month.

Mikhail Tyurin packs the docking probe in a stowage bag in Unity. The docking probe successfully guided the arrival of the Russian-built Pirs docking compartment to the International Space Station.   (NASA)

Mikhail Tyurin prepares the Russian Orlan space suit for a spacewalk.   (NASA)

First EVAs from Pirs

The Expedition 3 crew performed 4 EVAs from the Pirs Docking Compartment, inaugurating that facility for use through the early years of the ISS project. The Pirs EVAs used the Russian Orlan suits.

EVA-1 (ISS Russian EVA-2) was conducted on 8 October 2001. This EVA, the first from Pirs, was conducted by Vladimir Dezhurov and Mikhail Tyurin. It was also the 100th Soviet/Russian EVA.

The two Russians exited through the VL-1 hatch of Pirs (one of the two hatches on the module).

Equipment was stored during the EVA in the depressurised Pirs airlock and the node of the Zvezda module to which it was docked. The extra volume was needed for storage of the equipment which the cosmonauts would transfer outside. Culbertson remained in the Zarya/Unity/Destiny portion of the station so that Zvezda was sealed off (the Soyuz TM-32 ferry ship was docked to Zarya's nadir port and remained accessible).

The first tasks for the duo involved the connection of the Transit-B Orlan EVA suit communications system cable from the Pirs to Zvezda and the installation of handrails around the exit hatch and an access ladder.

Just over one hour into the EVA the two had removed thermal blankets from the Pirs and begun the installation of the first of two planned Stella booms (the GStM-1) to be used for future movements around the exterior of the complex.

The base of the Stella was installed first and the two men then set about attaching the extendable boom itself.

During periods of orbital night the cosmonauts took time off to rest and look at the dark Earth passing below. At one stage a report said the crew were taking the tasks at a "very leisurely" pace.

During the EVA Culbertson controlled the TV cameras of the Canadarm2 system and spectacular pictures were returned of Dezhurov and Tyurin working on the surface of the Pirs and also tossing the thermal covers from the Docking Compartment which they had removed into space.

Some 3 hours after their exit into space the cosmonauts began installation of antennas (AR-VKA and 2AR-VKA) for the Kurs automated rendezvous system to Pirs and inspected another of the Kurs antennas- 4AO-VKA.

As planners on Earth discussed extending the EVA, which was originally scheduled to last for 4 hours, the pair attached a docking target to Pirs. There were discussions with the ground regarding the actual positioning of the docking target which led to the two cosmonauts removing the target and reinstalling it.

With this delay and the general lengthening of the planned timeline the TsUP decided to postpone the final task of the EVA - a test of the operation of the Stella during which Dezhurov would have moved to the end of the extended boom - was postponed and the cosmonauts were told to return into the Pirs ("stop entertaining your-selves, just go back to the hatch," TsUP radioed).

The hatch closed after an EVA of 4 hours and 58 minutes.

The second EVA of Vladimir Dezhurov and Mikhail Tyurin (Russian ISS EVA-3) on 15 October was devoted to science rather than continued construction of the station.

Using the recently-installed ladder as a bridge between the Pirs and Zvezda, the cosmonauts crawled towards the rear of the module where they placed a detector containing sample plates called Kromka to monitor the amount of contaminant particles discharged around the thrusters of the module to allow for future thrusters to be better designed to protect the station's hull from the discharges.

The original plan had been to place the Kromka detector near thrusters on the Zenith plane of the Zvezda module but the cosmonauts had earlier noted a discolouration near that site and the planners had changed the location to the port side of the module.

Just before the two-hour mark of the EVA the Kromka was installed and the pair began the second part of their EVA plan, installing the Japanese Space Agency's MPAC&SEED experiments outside the Zvezda large diameter section.

The three "suitcase" size containers held a variety of samples for exposure to space including a micrometeorite detector (MPAC) which used gels and foams to capture tiny particles, and the SEED which included samples of paints, insulation and solid lubricants. The experiments were mounted on a load-bearing frame.

Once mounted the experiment packages were opened and photographed before being left where they were expected to remain exposed to open space for 1 to 3 years.

The final activities of Dezhurov and Tyurin involved recovery of an exposed Russian flag with special paints and its replacement with cards containing the logo of a photographic company as part of a US-Russian commercial agreement to provide external cameras for future documentation of ISS construction activities both for the ISS controllers and through an Internet web site.

Dezhurov and Tyurin photographed the placards before they returned to the Pirs airlock to wrap up a 5 hour 52 minute EVA.

Redocking Soyuz TM-32

On 19 October 2001 the three men of Expedition 3 entered the Soyuz TM-32 ferry ship, sealed themselves in and undocked from the Zarya FGB nadir port.

Dezhurov flew the craft along the length of the FGB module until the craft arrived at a position in front of the Pirs docking unit. He then flew TM-32 to a docking with that module.

Taxi Mission 2

Soyuz TM-33 was launched on 21 October 2001 to conduct the second planned swap of the Soyuz return craft at the ISS.

The two Russians of the crew were Viktor Afanasyev, a Mir veteran commander and rookie Konstantin Kozeyev. They were accompanied by French Researcher, and ESA astronaut, flying a science mission on behalf of the French Space Agency CNES, Claudie Haignere.

Soyuz TM-33 was docked to the Zarya nadir port on 23 October and the two crews of Expedition 3 and the Taxi 2 mission conducted a week of scientific research and other activities.

Afanasyev, Kozeev and Haignere returned to Earth on 31 October in Soyuz TM-32 leaving the fresher TM-33 craft for the Expedition 3 crew.

Frank Culbertson wearing thermal undergarment, adjusting his communication headgear in Zvezda.   (NASA)

First EVA for Culbertson

The third EVA (Russian ISS EVA-4) was on 12 November featured Valdimir Dezhurov and was the first and so far only excursion outside for Frank Culbertson. The pair

- installed handrails on the Pirs module and the transition piece of Zvezda;
- laid a cable, Kurs- P VCH, from Zvezda to Pirs, joining the electrical disconnects;
- laid a ribbon cable from Zvezda to Pirs and connected it to the Kurs- P NCH antenna;
- fixed cables on fastening elements installed earlier;
- tested the extension of the Stella cargo crane (GStM-1);
- inspected and photographed a small section of a solar battery, SB-II, on the Zvezda module which had failed to deploy after launch on 12 July 2000 which had remained stuck ever since.

It took Culbertson and Dezhurov about 2 hours to route the first bundle of high frequency cables from the Zvezda where they were originally located and routed them down the side of the Pirs compartment where they were connected to a panel.

The second cable bundle was then routed and connected, replacing an older low frequency set, some 30 minutes after the first set of cable connections.

Culbertson and Dezhurov then moved over to the Zvezda port side, installed some foot restraints and took documentary pictures of the small section of undeployed solar panel.

During the periods of orbital darkness the pair took time to observe the sights of the Earth below. Culbertson remarked about seeing the stars and lights from cities in the Middle East.

He remarked to Tyurin, who was inside the complex, that all he needed now was music (Culbertson, a keen Jazz fan, had a trumpet aboard the complex and relaxed by playing it.)

Once back in orbital daylight Dezhurov extended the Stella boom and manoeuvred it in pitch and yaw.

Once the boom was fully extended to around 10 metres Dezhurov began winding the crank handle some 80 times to retract it. A further test, during which Culbertson was to translate down the crane, was cancelled by controllers, reportedly because the initial tests went so well.

Happy with their efforts the pair cleaned up some tethers and moved back into the Pirs module and closed the hatch after a 5 hour 4 minute excursion.

New Progress Could not Complete Docking

Progress M1-7 was launched on 26 November and the Progress M-45 craft undocked the next day.

The docking of Progress M1-7 could not be completed on 28 November as planned because of a foreign object in the docking cone of the Zvezda which required and EVA by the two Russians of the crew to remove (see the "Emergemcy EVA" section below) before the final docking could be accomplished on 3 December.

Emergency EVA

The fourth excursion outside the complex (Russian ISS EVA-5) was the first emergency EVA of the ISS programme and was conducted on 3 December to remove the obstruction which had prevented the docking of the Progress M1-7 craft to the rear port of Zvezda.

The EVA was conducted by Dezhurov and Tyurin.

Just over an hour after the hatch of Pirs was opened the pair had reached the aft plane of the module and reported that the object was a rubber sealing ring from the active Progress M-45 docking unit (left behind as that craft undocked) which was caught around the docking port and was twisted in places.

Ground controllers at TsUP commanded the Progress docking probe to extend giving some 395 mm of clearance for the cosmonauts to insert a cutting tool that they had fashioned from items inside the station and cut the sealing ring in two and remove it for return into the complex.

TsUP then commanded the Progress to retract the docking probe and engage the hooks and latches contained on the docking port faces to latch. At 1454 TsUP reported that a hard dock had been achieved completing the docking of the Progress M1-7 craft with the ISS.

Dezhurov and Tyurin then took photographs of the area and of science instruments placed outside earlier before beginning the trek back towards the Pirs compartment.

The pair closed the hatch of the Pirs module at 1606 completing their "labour intensive" EVA giving an official duration of 2 hours 45 minutes and 30 seconds.

Energia said that the Progress M-45 seal loss event was the first time in 30 years of use of the design of the docking probe that such a problem had happened.

The emergency EVA delayed the launch of the Expedition3's replacements by a few days, the only extension to the mission of the 3rd crew.

Expedition 3 Ends

The Expedition 4 crew was launched on the Endeavour Orbiter late on 5 December and docked with the ISS complex two days later delivering another MPLM – Raffaello – containing tonnes of equipment and supplies including new science experiments and equipment.

The seatliners of the Expedition 3 crew were swapped with those of the Expedition 4 crew (Yuri Onufrienko, Carl Walz and Dan Bursch) on 9 December and the Expedition 3's tenure on the ISS was officially at an end.

During the handover period the crews worked to replace major components of the TVIS treadmill and the replacement by Onufrienko and Usachev of a compressor in the secondary redundant loop of the Zvezda air conditioning system. The primary system had been leaking Freon at a very small rate and the redundant loop needed to be activated so that the primary system could be repaired later.

Before the end of their flight, however, the crews of Expedition 3 and 4 and the shuttle crew paid tributes to those killed during September's terrorist atrocities.

Endeavour undocked on 15 December leaving Expedition 4 in charge of the complex.

The craft landed on 17 December completing the shortest Expedition to date on the ISS and closing out the first year of continuous operation of the complex.

Flight Events of Expedition 3 at a Glance

Duration (launch to landing): 128 days 20 hours 44 minutes and 56 sec

Date	Event
August 2001	
10/8	launch of STS-105 Discovery carrying Expedition 3)
12/8	docking of Discovery STS-105
20/8	undocking of STS-105
21/8	launch of Progress M-45
22/8	undock progress M1-6, destruction in the atmosphere
23/8	docking of Progress M-45 Zvezda rear port
September 2001	
14/9	launch of Progress M-SO-1
17/9	docking with Zvezda nadir of Progress M-SO-1
26/9	undocking of Progress from SO-1 (Pirs), deorbit
October 2001	
8/10	EVA-2 from Pirs (Dezhurov, Tyurin) 4:58
15/10	EVA-3 from Pirs (Dezhurov, Tyurin) 5:51
19/10	redock of Soyuz TM-32 from FGB nadir to Pirs (Zvezda nadir)
21/10	launch of Soyuz taxi-2 (TM-33)
23/10	docking of Soyuz TM-33 to FGB nadir
31/10	undock of Soyuz TM-32 with Taxi-2 crew from Pirs
November 2001	
12/11	EVA-4 from Pirs (Dezhurov, Culbertson) 5:05
22/11	undocking of Progress M-45 from Zvezda rear, deorbit
26/11	launch of Progress M1-7
28/11	soft dock of PM1-7 with rear Zvezda port
December 2001	
3/12	EVA -5 to removed obstruction PM1-7 docks with Zvezda port 2:46
5/12	launch of Endeavour STS-108 (Expedition 4)
7/12	STS-108 docks
15/12	undocking of STS-108
17/12	landing of STS-108 and the Expedition 3 crew

ISS at One Year

Observations by Frank Culbertson

At the end of Expedition 3 Frank Culbertson was debriefed by NASA and provided information about the current habitability status of the complex.

He said that the three most important issues facing the ISS programme at that time were: "time/information management, cargo stowage, and communications…with degraded or lost communications representing the most serious concern."

During one of the first transfer operations with the MPLM containers the crew noted small metal particles floating in the module after it was opened, the ground crews addressed this problem for the next flights and Culbertson reported that, although the crew wore goggles and rubber gloves when opening either the MPLM or the Progress craft, they were very clean.

Frank L. Culbertson Jr, takes a brief timeout from a busy day to play his trumpet in the Quest Airlock. **(NASA)**

The Expedition 3 Commander said that the crew could currently cope with the demands of the stowage of various cargoes although much time was expended on this activity and keeping the Inventory management system updated.

Stowed items were generally not a hindrance to working on Expedition 3 but concern was expressed about the ability to gain access to panels and fire extinguishers in the future.

As more science racks arrived and replaced stowage racks the room for stage would be reduced. Culbertson recommended an approach to discard the older items on the station if they had not been used and also for the retention of an MPLM docked to the ISS between shuttle missions for extra stowage space.

Culbertson also noted a slight temporary hearing loss because of the noise in the Russian Segment and the Zarya module, although he stressed that this was not a major concern and did not affect their working. The crew wore foam ear plugs or headsets when working with noisier equipment.

Air quality in the complex was "very good" Culbertson reported with all of the filters and ventilation systems operating "well". However, Culbertson noted difficulty in accessing certain filters.

Culbertson noted that although the crew was aware of the Freon leak from the Zvezda air conditioning unit no odour was detected from the leak.

Condensation was not a problem on his mission, Culbertson reported, although large condensate "blobs" had been found on the air conditioning loop during the first expedition.

Movement inside the complex was easy but there were many handrails in the US Segment and elsewhere, which were "head and knee knockers" although he noted that they were of more concern for the visiting expeditions rather than the expedition crews.

Other items, such as needles used by the Expedition crews for a variety of purposes also needed care by the visiting crews, Culbertson noted.

Regarding pollution around the exterior of the complex, Culbertson said that pollution from the Russian thrusters was not a problem (although the Russian side had placed the Kromka witness plate near to Zvezda jets to record the deposits emitted during these firings).

For EVAs all crews had drawings to show "translation paths" through the maze of thruster packages and antennae on the modules so that designated work sites could be reached.

Following his EVA, Culbertson noted that there was a "smell of space" (attributed to atomic oxygen) inside the complex after the EVAs but no thrusters or contamination odours were detected. Already, Culbertson said, the leading edges of the complex were showing tan stains due to interaction with environmental dust and atomic oxygen.

Source: *"Expedition III Crew Debriefing - Minutes of the Meeting - January 31, 2002" Posted on the "Spaceref.com" and "NASAwatch.com" web sites*

NASA Shuttle Missions to ISS (1998-2001)

DAVID J. SHAYLER

Introduction

On 10 August 1968, less than a year before Apollo 11 accomplished the first manned landing on the moon, Dr. George E. Mueller, the Associate Administrator for Manned Spaceflight at NASA, in his address to the British Interplanetary Society held at University College London in honour of his election to Honorary Fellow, forecast that, "The next major thrust in space will be the development of an economical launch vehicle for shuttling between Earth and installations, such as the orbiting space stations, which will soon be operating in space. Essential to the continuous operation of the space station will be the ability to re-supply expendables as well as change and/or augment crews and laboratory equipment. Therefore, there is an early requirement for an efficient Earth-to-orbit transportation system – an economical space shuttle...(which) according to current studies could carry anywhere between 25,000 and 50,000 lbs of payload that could be stored in cargo compartments within the central vehicle. (The) design of space stations and payloads is presently under study within NASA. The maturity of these designs would coincide with those for the space shuttle" [1].

During the late 1960s, as part of post-Apollo planning for the next phase of American exploration of space, studies had identified the need for a reliable and affordable space transportation infrastructure that encompassed not only routine access to Earth orbit, but also a network of lunar research bases and the beginning of human expeditions to Mars. The focal point of this plan was the creation of a permanent Earth orbital space station, re-supplied by a reusable logistics re-supply craft capable of shuttling crews and cargo from Earth to space and back again. By 1972, after severe budget restrictions, all that remained of this grand plan was the Space Shuttle system, which struggled to answer the needs of the American space programme over the next two decades. It was not until the 1984 Presidential commitment to finally build a space station that the Shuttle was allowed to achieve the primary objective it had originally been designed for, but not without further delays in choosing the final configuration of the space station. After successfully demonstrating its potential during ten visits to space station Mir between 1995-1997 (also called Phase 1 see [2]), NASA's fleet of shuttle orbiters have become the workhorse of the current space station programme, transporting hardware elements, crews and supplies to and from orbit during the assembly phase that created ISS.

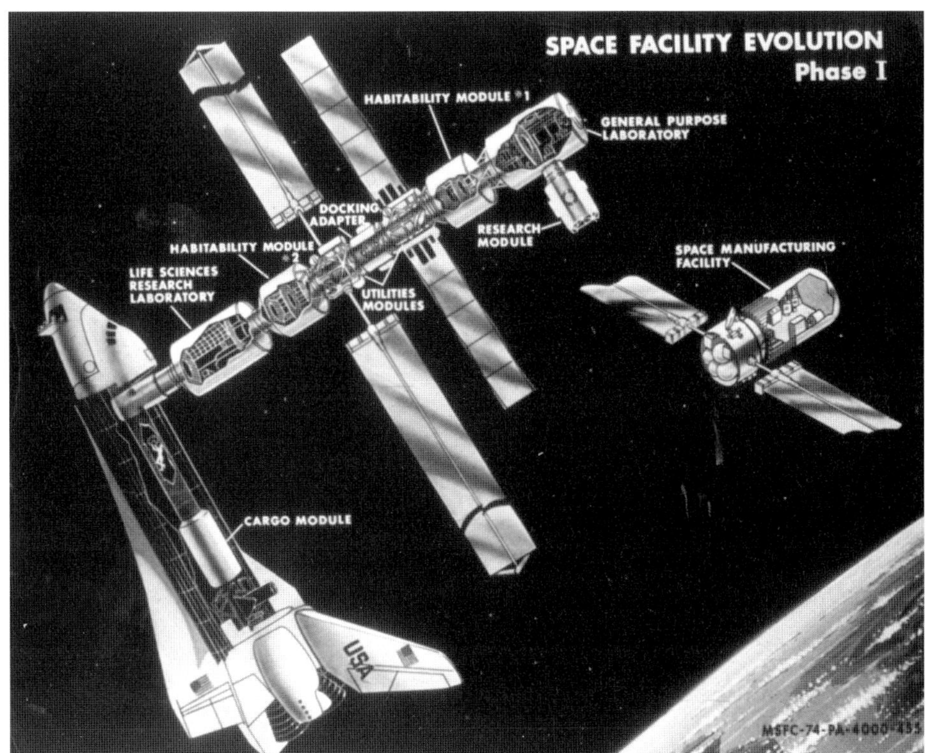

A 1974 artists impression of a shuttle docking with a space station. 25 years later this has evolved into the ISS (see front cover).   (NASA)

The Missions

It would be thirty years after Mueller delivered his speech before NASA would finally be ready to dispatch the first Shuttle mission in the ISS assembly manifest. The launch of STS-88 (2A), in December 1998 was the first of 12 Shuttle launches over the next three years that encompassed Phase II (launch of first elements to the point where ISS was ready to begin scientific research in the US Lab, Destiny) and early Phase III (gradual progression towards completion of the station assembly) of the ISS programme. Although each of these 12 missions were distinctly different, they all followed a basic flight profile that can be broken down into elements, reviewed in this paper, which illustrates the role the Shuttle system has in supporting ISS operations independent of the activities of the resident crews and the participation of the other international partners.

One of the American shuttle orbiters arrives at ISS. Visible in the open payload bay is the forward mounted docking facility and the latest elements for installation on the station during the mission. As the primary transportation vehicle for initial assembly and outfitting the Shuttle will continue to play a prominent role during the expected life of the station.   (NASA)

Of the 12 missions, Atlantis (OV-104), Discovery (OV-103) and Endeavour (OV-105) completed four missions each. Columbia (OV-102) was too heavy to reach the new inclination with the added weight of the early station hardware and would instead serve as the orbiter to support non-ISS missions (such as the Hubble Service Missions) during the 1998-2002 period.

Planning for Flight

Mission planning has to take into account the availability of flight hardware. This includes the shuttle orbiters themselves, which have to complete regular upgrade and maintenance programmes of several months duration in addition to normal mission processing – thus periodically taking the vehicle out of service. The inventory of Shuttle flight hardware also includes scheduling the SSME's, ET, SRB segments, RMS, EVA equipment, cargo carriers, crew equipment and provisions, and fuel. The ground support infrastructure includes organising hardware, payload processing and countdown sequence, as well as scheduling the global tracking and recovery teams, various management, administrative, training and mission control teams and their support staff and, of course, the assigned flight crew. Scheduling of the ISS elements, secondary payloads and experiments becomes an integral part of each mission early in the planning and is also dependent upon the availability of hardware supplied from US and International contractors, the current status of the shuttle fleet, ISS on-orbit condition and budget restraints. The consequences of the timing of various elements became very evident in the delay to orbital construction during 1999-2000, with the difficulties encountered in the shuttle electrical subsystem that affected the whole fleet, and preparing the Russian Service Module for launch.

In contrast, operations from mid 2000 to the end of 2001 progressed more smoothly because the US and Russia delivered payloads to orbit on time. As one NASA official pointed out several years before the start of flight operations, the sequence of early ISS assembly missions – at least to the delivery of the US Laboratory – depended

TABLE 1: *STS-ISS Missions Summary 1998-2001.*

STS Flight	ISS Mission	Shuttle Orbiter	Crew Size	Launch Date	Launch Time (EST)	Pad 39	Landing Date	Landing Time (EST)	Landing Orbit	Duration to Wheelstop dd:hh:mm:ss	Miles (millions)
88	2A	OV-105	6	1998 04 Dec	0335	A	1998 15 Dec	2253	185	11:19:18:47	4.7
96	2A.1	OV-103	7	1999 27 May	0649	B	1999 06 Jun	0202	154	09:19:13:57	3.8
101	2A.2a	OV-104	7	2000 19 May	0611	A	2000 29 May	0220	155	09:21:10:10	4.1
106	2A.2b	OV-104	7	2000 08 Sep	0845	B	2000 19 Sep	0356	185	11:19:12:15	4.9
92	3A	OV-103	7	2000 11 Oct	1917	A	2000 24 Oct	1459	203	12:21:43:47	5.3
97	4A	OV-105	5	2000 30 Nov	2206	B	2000 11 Dec	1804	170	10:19:58:20	4.5
98	5A	OV-104	5	2001 07 Feb	1831	A	2001 20 Feb	1533	202	12:21:21:00	5.3
102	5A.1	OV-103	7	2001 08 Mar	0642	B	2001 21 Mar	0231	201	12:19:51:57	5.3
100	6A	OV-105	7	2001 19 Apr	1441	A	2001 01 May	1211	186	11:21:31:14	4.9
104	7A	OV-104	5	2001 12 Jul	0504	B	2001 31 Jul	2339	200	12:18:36:39	5.3
105	7A.1	OV-103	7	2001 10 Aug	1710	A	2001 22 Aug	1423	186	11:21:13:52	4.3
108	UF-1	OV-105	7	2001 05 Dec	1719	B	2001 17 Dec	1255	185	11:19:36:45	4.8

Notes: ISS (3 person) crew transfers: ISS-1 launched on Soyuz TM-31 returned on STS-102; ISS-2 up on STS-102 down on STS-105; ISS-3 up on STS-105 down on STS-108; ISS-4 up on STS-108 to be returned on 112 in 2002. EST - Eastern Standard Time.

upon the very next launch to maintain the momentum. Likened to standing up a row of dominos, if one fell it would have a serious knock on effect down the line. This was clearly demonstrated in 1999 with the delays to Zvezda and again in 2002 with faults in the Shuttle main engines, and will remain a major influence in annual ISS operations.

Instrumental to the overall success of Shuttle operations in the 1998-2001 period was the creation of the Space Flight Operations Contract, initiated by NASA in August 1995, to progress to a single prime contractor for Space Shuttle operations. The same month, Rockwell and Lockheed Martin agreed to form a new joint space operations venture under the name United Space Alliance (USA). In April 1996, the management of Rockwell Space Operations Contract (SOC) and Lockheed Martin Shuttle Processing Contract (SPC) trans-ferred to USA, with the Space Flight Operations Contract coming into effect in October 1996. The first mission managed by USA was STS-80 in November 1996. Under SFOC guidelines, USA would become responsible for the following Shuttle operations:

Space Hardware Processing: including planning and integration of payloads; processing of the shuttle vehicle; management of logistics and supply chains; operation and management of facilities at the Cape; and management and verification of the final launch configuration.

On-Orbit Operations: simulation and execution of mission planning forecasts; planning and control of the Flight Equipment Inventory; contingency planning and execution; and control and management of orbital systems.

Launch and Return Operations: design of each flight and its integrated mission planning; contingency planning and operations; development and integration of all software; real time systems and control centre (MCC-H) operations.

Space Systems Training: training and certification of space systems operators (astronauts); launch and flight controller team training and certification; hardware processing training and certification; and management of all simulation and training facilities.

The primary objective of USA was to maintain safety and reliability as its top priorities, while at the same time reducing the overheads of operating the shuttle fleet in areas of flight preparations, on-time launches and safe landings. At the same time, there was a significant reduction in the overall shuttle budget of $300 million during the period of FY1996 to FY1998. In the first five years of SFOC operations (1996-2001), which encompassed the conclusion of Shuttle Mir and initial ISS operations, USA claimed that the shuttle was now flying more safely, more economically and more reliably than at any other time in the program's 20 years history. The company also cited that the average number of Shuttle In-Flight Anomalies (IFAs) had dropped from the average of 20 between STS-26 and STS-79 (from the peak of 61 for STS-26), to an average of 6 between STS-80 and STS-109 (peaking at 12 IFAs on STS-89 and STS-92). The USA handouts also proclaimed a 100% mission success in 29 missions flown safely and successfully; a 67% reduction in launch scrubs for technical reasons; and an 83% launch on-time (excluding weather delays) [3].

This certainly helped to ensure that the Shuttle was ready to deliver ISS components and logistics on time and safely, maintaining the momentum of initial construction and sustaining orbital operations with the resident crews to the recovery of the fourth resident crew in June 2002.

Shuttle Mission Control Centre – Houston (MCC-H)

Overall authority for Station Operations is retained by Mission Control Center in Houston (MCC-H) for all phases of the programme, sharing the responsibly with Mission Control Center Moscow (MCC-M) for vehicle control functions for the respective American and Russian segments of the station. Each centre is capable of providing a backup role to the other for critical functions if required. Prior to Flight 5A, MCC-M was responsible for the execution of leadership decisions, while MCC-H became the leading centre for the 'execution of multisegment procedures' that required interface or interaction between the US and Russian segments, as well as overall leadership. From the beginning of the construction phase of ISS, MCC-H also retained overall leadership of all US space shuttle flights to ISS from a separate control room in Building 30 at JSC.

A wide-angle view from the rear of the new Shuttle mission control facility located at the Johnson Space Centre near Houston. This photo was taken during the first operational   use of the station during STS-70 in July 1995. (NASA)

The original Mission Control Center in Building 30 at JSC was decommissioned in 1996 after 30 years service. It was designated a national historical landmark and is now called the Apollo Mission Control Center. The current MCC is located in the new adjacent facility (Building 30 South) and is used to support all Shuttle/Space Station missions and simulations. There are currently two different types of Flight Control Room (FCR pronounced 'Flicker') in Mission Control. The Space Shuttle Flight Control Room (the White Room) is used for all Shuttle operations – whether at ISS or for independent flights such as the Hubble Service Missions – while the ISS Flight Control Room (the Blue Room) is used to support of ISS increment crews. Both are basically identical in the equipment provided and the support structure, but the Shuttle flight control room is larger in volume and has more controllers than the ISS room, although data can be transmitted and viewed in either room.

The Shuttle control team is divided into teams (Ascent/Entry – Orbit – Planning) who occupy the consoles on a rotational basis 24 hours a day for the duration of the mission. When the shuttle crew sleep, an MCC 'night shift' is on duty; it is also known as the 'graveyard' shift as it is normally a quiet tour. Each MCC team trains on various simulations of the mission in the months leading up to launch and these can include Integrated Sims with the assigned flight crew in the Shuttle simulators. Several members of the control teams can work on several upcoming missions at the same time, and in some cases have completed elements of astronaut training (such as RMS ops and EVA Sims in the WETF) to familiarise themselves with tasks assigned to the flight crews.

Crew Assignments and Training

The mission training of a Shuttle crew normally takes about 12 months, depending upon the flight plan, the previous experience of the crew and the current launch manifest. It was obvious that preparations for the first ISS missions would involve additional training, especially in rendezvous and docking and EVA assembly tasks, and that some sessions would be required in Russia, adding to the complexity of an already busy 'normal' shuttle mission training cycle.

TABLE 2: *MCC-H STS Console Manning.*

```
               =============================================
Row 1          |      /—1—V—2—\              /—3—\           |
(The Trench)    |                                            |
Row 2          | /—4—V—5—\            /—-6—V—7——\    |
               |                                            |
Row 3          | /—8—V—9—\            /—10—V—11——\   |
               |                                            |
Row 4          | /—-12——\        /—13—/\—14—\  /—15—\  |
               |                                            |
Row 5          | /—16—\       /—-17—/\—18—\     /—19—\ |
```

Console	Controller	Initials	Callsign
1	Flight Dynamics Offcier	FDO	'Fido'

Defines shuttle performance capability during the ascent. Also plans and targets all orbital translational manoeuvres including the critical end of mission deorbit burn

Console	Controller	Initials	Callsign
2	Rendezvous, Guidance & Procedures Officer	GDO	'Guido'

Ensures the onboard navigational and guidance computer software completes their tasks to meet mission objectives

Console	Controller	Initials	Callsign
3	Ground Controller	GC	'GC'

Director of all maintenance and operation activities affecting MCC hardware, software and support facilities. Coordinator of spaceflight tracking and data network including the TDRS satellites with Goddard

Console	Controller	Initials	Callsign
4	Propulsion	PROP	'Prop'

Monitors & evaluates RCS and OMS systems and manages levels of propellant and other consumable for manoeuvring

Console	Controller	Initials	Callsign
5	Guidance, Navigation & Control Systems	GNC	'GNC'

Monitors all vehicles guidance, navigation and control systems and informs FD & crew of impending abort situations, and advised flight crew of guidance malfunctions

Console	Controller	Initials	Callsign
6	Maintenance, Mechanical Crew Systems Engineer	MMACS	'Max'

Monitors orbiter structural and mechanical systems and follows use of onboard crew hardware and in-flight maintenance equip.

Console	Controller	Initials	Callsign
7	Electrical Generation & Integrated Lighting System Engineer	EGIL	'Eagle'

Monitors cryogenic levels on fuel cells, electrical generation and distribution systems and the vehicle lighting

Console	Controller	Initials	Callsign
8	Data Processing System	DPS	'DPS'

Determines the status of the data processing system which includes the 5 onboard GPC's, flight critical and launch data lines, the displays, onboard mass memory and software

Console	Controller	Initials	Callsign
9	Payloads Offcier	PAYLOADS	'Payloads'

Coordinator of the onboard and ground system interfaces between the FAO and payload user, monitors Spacehab and carrier support systems and interfaces with the payload

Console	Controller	Initials	Callsign
10	Flight Activity Officer	FAO	'FAO'

Coordinator for planning and supporting onboard crew activities, checklists, procedures and schedules

Console	Controller	Initials	Callsign
11	Emergency Environmental & Consumable Systems Engineer	EECOM	'EECOM'

Responsible for the monitoring of cooling systems in the cabin and avionics compartment and the cabin pressure control systems

Console	Controller	Initials	Callsign
12	Instrumentation & Communication Offcier	INCO	'INCO'

Planning and monitoring the configuration of all in-flight communication and instrumentation systems

Console	Controller	Initials	Callsign
13	Flight Director	FD	'Flight'

Leader of the Flight Control Team (FCT) – the 'boss'. Overall responsibility for the shuttle mission and payload operations. The FD is also responsible for all decisions regarding the conduction of a safe and expedition flight

Console	Controller	Initials	Callsign
14	Spacecraft Communicator	Capcom	

An astronaut who serves as a primary voice contact between the FCT and flight crew.

Console	Controller	Initials	Callsign
15	Payload Deployment Retrieval Systems	PDRS	'PDRS'

Responsible for monitoring all operations with the RMS

Console	Controller	Initials	Callsign
16	Public Affairs Officer	PAO	'PAO'

Supplies on going mission commentary to supplement and explain the air-to-ground transmissions to both the news media and public

Console	Controller	Initials	Callsign
17	Mission Operations Director	MOD	'MOD'

Member of the Management team at JSC providing a representation and linked to NASA management structure in other centres and Mission Management Teams

TABLE 2: *MCC-H STS Console Manning (Contd).*

Console	Controller	Initials	Callsign
18	Russian Liaison Officer/ Booster Systems/EVA Systems Engineer	RLO/Booster/EVA	

A rotational console depending on the requirement at the time of the mission.' Booster' monitors and evaluate the performance of the SSME/ SRBs/ET during pre-launch and ascent; 'RLO' serves as a point of contact between MCC-H and MCC-M; 'EVA' monitors the EVA crew and their pressure garments during EVAs

19	Flight Surgeon	SURGEON	'Surgeon'

Responsible for the monitoring of crew activities, coordinate the medical operations flight control team, serves as a consultation with the crew and advisor to the FD on aspects of crew health.

The prime crew conduct one of several press briefing held prior to the mission where they explain their objectives, flight plan and roles for the up coming mission as well as answering media questions. (NASA)

Availability of simulators and crew scheduling plays an important part in crew training. Upon the selection of the crew to a mission, each member is assigned particular responsibilities in addition to those normally associated with the role they have been selected for. Normally the Commander has the overall responsibility for mission success and crew safety, and on ISS missions is the primary lead for the ISS rendezvous and docking phase. The Pilot assists the Commander in the ascent and entry phases and as back up for docking to ISS. It has become a custom for the Pilot to perform the undocking and fly around manoeuvre at the end of joint operations.

Mission Specialists are normally responsible for RMS activities, EVA operations, transfer of logistics and operation of experiments carried. However, the increasing complexity of each mission requires dual responsibility and assignment in many areas, with Commanders serving as RMS operators and Pilots active in the role of back up EVA astronauts. On every Shuttle mission there is a long list of tasks to be completed by the crew prior to flight. Each task is assigned (by crew agreement under direction of the Commander) to an individual astronaut as the primary point of contact, with a second or third as back up to ensure cross training. These items vary from major payload elements, critical items of hardware or flight operations, to organising the official crew photo or designing the mission emblem.

With 6-8 crews in various states of mission training at any one time, it is clear that scheduling simulator time is dependent upon the flight sequence and the complexity of the mission (such as EVAs). Astronauts are also assigned to mission support roles on other missions (KSC launch support, Capcom, WETF support, SAIL etc.) prior to and in addition to their own specific mission training, and this helps maintain their proficiency. Beginning in 2001, several astronauts began to be recycled back for a second mission to ISS, drawing upon their past experiences. In the future, astronaut training will probably be focused on a smaller core of Shuttle 'orbiter' crews (Commander, Pilot, MS1, MS2), who may fly several shuttle re-supply missions together to reduce training time and costs and help lighten the burden on limited simulator time.

STS-88 astronauts participate with training staff from USA and JSC in the Crew Equipment Interface Test (CITE) held at the KSC in Florida. Here the flight crews familiarise themselves with crew equipment they are to use during the mission and which is stowed in the orbiter payload bay and crew compartments.

(NASA)

TABLE 3: *STS Flight Crew Positions for ISS Construction Missions.*

Assembly Flight	STS	OV	CDR	PLT	MS1	MS2/FE	MS3	MS4	MS5
ISS-01-2A	88	105	Cabana	Sturckow	Ross	Currie	Newman	Krikalev[1]	-
ISS-02-2A.1	96	103	Rominger	Husband	Jernigan	Ochoa	Barry	Payette[2]	Tokarev[1]
ISS-03-2A.2a	101	104	Halsell	Horowitz	Weber	Williams J.	Voss J.S.	Helms	Usachev[1]
ISS-04-2A.2b	106	104	Wilcutt	Altman	Lu	Mastracchio	Burbank	Malenchenko[1]	Morukov[1]
ISS-05-3A	92	103	Duffy	Melroy	Chiao	McArthur	Wisoff	Lopez-Alegria	Wakata[4]
ISS-06-4A	97	105	Jett	Bloomfield	Tanner	Garneau[2]	Noriega	-	-
ISS-07-5A	98	104	Cockrell	Polansky	Curbeam	Ivins	Jones	-	-
ISS-08-5A.1	102	103	Wetherbee	Kelly J.	Thomas A.	Richards P.	-	-	-
ISS-09-6A	100	105	Rominger	Ashby	Hadfield[3]	Philips	Parazynski	Guidoni[3]	Lonchankov[1]
ISS-10-7A	104	104	Lindsey	Hobaugh	Gernhardt	Kavandi	Reilly	-	-
ISS-11-7A.1	105	103	Horowitz	Sturckow	Forrester	Barry	-	-	-
ISS-12-UF1	108	105	Gorie	Kelly M.	Godwin	Tani	-	-	-

Key: OV-Orbiter; CDR-Commander; PLT-Pilot; MS – Mission Specialist; FE –Flight Engineer (ascent/entry); 1. Russian Space Agency; 2. Canadian Space Agency; 3. European Space Agency;4. Japanese Space Agency.

Notes: ISS (3 person) crew transfers: ISS-1 launched on Soyuz TM-31 returned on STS-102; ISS-2 up on STS-102 down on STS-105; ISS-3 up on STS-105 down on STS-108; ISS-4 up on STS-108 to be returned on 112 in 2002.

STS-88 (2A): **On 16 August 1996, NASA announced the five-person Shuttle crew for the first ISS assembly flight (NASA News Release - 96-169) – then scheduled for late 1997 but ultimately delayed until December 1988. Then, on 30 July 1998 (98-137), veteran cosmonaut Sergei Krikalev, already in training as a member of the first increment crew, was added to the STS-88 crew. NASA stated that Krikalev's previous experience on Mir (two missions totalling 15 months and 7 EVAs) combined with his experience on the STS-60 crew (during which he operated the RMS) and his support role during Shuttle-Mir offered unique experiences for this important and historic mission.**

STS-92 (3A): **On 2 June 1997 (98-117), NASA announced that Japanese astronaut Koichi Wakata would be assigned as primary RMS operator for STS-92, due to his previous RMS experience on STS-72 where he retrieved the Japanese Space Flyer Unit. As NASA Administrator Goldin explained; "Mr. Wakata's... participation in such an early and significant space station assembly mission...is symbolic of the close bond that has developed (between the US and Japanese programmes) and the extent to which we rely on one another (on ISS)." A week later on 9 June (97-126), the other four (EVA) MS were officially assigned to the flight. It was not until 3 February 1998 (98-21) that Commander Brian Duffy and Pilot Pam Melroy were assigned to complete the crew. The fact that Chiao, Duffy and Wakata had flown together on STS-72 would certainly help during the training for the mission.**

On 9 June 1997 (97-126), NASA released the names of a cadre of 14 astronauts who would be trained in the extensive programme of on-orbit construction for initial ISS flights, expected to be completed over six missions and several months of flight operations. The importance of an early start in crew training was underlined by David Leestma, Director of Flight Crew Operations and former astronaut, himself experienced at EVA, who commented; "These crewmembers will be exceptionally busy preparing for some challenging and demanding tasks, from initial assembly, through installation of the robotic arm (Canadarm2) and an airlock (Quest) for station based space walks." It was expected that these assignments would be refined, with additional tasks assigned as mission requirements demanded. STS-88 (2A) astronauts Jerry Ross and Jim Newman had already been in training for some months and would now be joined for the expanded EVA training programme by:

STS-92 (3A): Leroy Chiao; Jeff Wisoff, Michel Lopez-Alegria, Bill McArthur (all named earlier that month)

STS-97 (4A): Joe Tanner; Carlos Noriega

STS-98 (5A): Mark Lee and Tom Jones

STS-99 (6A): Chris Hadfield and Robert Curbeam

STS-100 (7A): Mike Gernhardt and James Reilly

STS-96 (2A.1), 97 (4A) & 98 (5A): A total of thirteen crewmembers were assigned to these three missions on 4 August 1998 (98-143), including Yuri Malenchenko as part of the STS-96 crew, which was to have been launched after the Russian Service Module, Zvezda in April 1999. Malenchenko's training had specialised in the preparation of Zvezda and unloading of the Progress prior to the arrival of the first resident crew.

STS-101 (2A.2a): On 16 November 1998 (98-205), a team of five astronauts were named to the third mission to fly to ISS. Originally it was planned to launch this mission after the delivery of Zvezda during the late summer of 1999, but when that became delayed, a change to the Shuttle manifest also required a change in the make up of the crew. Three months later, on 12 February 1999 (99-19), three Russian cosmonauts were assigned to upcoming Shuttle flights to give additional experience and expertise. Valery Tokarev replaced Yuri Malenchenko on STS-96 because of the latter's more focused training. With Zvezda delayed, Malenchenko was now reassigned with Dr Boris Morukov (who was unsuccessful in a Russian bid to be assigned to STS-88) on STS-101, which would now fly *before* Zvezda launched to continue outfitting the Functional Cargo Block (FGB – Zarya) with additional supplies while awaiting the Zvezda launch.

By the end of 1999, the ISS (Zarya/Unity) had been in space for a year, visited by only two shuttle crews. Zvezda was likely to be delayed until July 2000 and the Shuttle fleet was undergoing electrical wiring inspections, further delaying flight operations. Therefore, on 18 February 2000 (00-29), after a review of available options, NASA split the objectives of STS-101 into two missions. Originally, thought was given to flying 101 before Zvezda launched and then a reflight (101-A) with same orbiter and crew after Zvezda had docked to ISS. But the agency decided against this. Instead, a new mission was inserted into the manifest (STS-106, the next available number) and Lu, Malenchenko and Morukov were moved to the new mission. They were replaced on STS-101 by the Increment 2 crew (Usachev, Voss and Helms), who would help with onboard preparations the for Increment 1 crew and would gain valuable experience in working aboard the station that would help them quickly adjust during the change over with ISS-1. Commander Wilcutt, Pilot Altman, and MS Burbank and Mastracchio were named with Lu, Malenchenko and Morukov to *STS-106 (2A.2b)* on 18 February 2000.

STS-102 (5A.1): The names of the STS-102 crew were released on 9 May 2000, and those for *STS-100 (6A)* and *STS-104 (7A)* on 28 September 2000 (00-154). STS-104 EVA astronauts Mike Gernhardt and James Reilly had been in training for their mission for over three years prior this announcement. The *STS-105 (7A.1)* crew was named on 1 December 2000 and the *STS-108 (UF1)* crew, the final announcement of crew assignments in this period, on 29 January 2001.

In all, 68 Shuttle crew positions were filled on the 12 missions, with an additional 9 seats taken up on three ISS resident crew exchanges. A total of 64 astronauts flew one mission and 4 (Rominger, Sturckow, Horowitz and Barry) flew twice. Four Shuttle crewmembers from early assembly flights (Krikalev, Voss JS, Helms and Usachev) all later flew as members of resident ISS crews.

Pre-launch Activities

By 20 November 1998 and the launch of the first ISS element – Zarya, the Russian Control Module – the American Space Shuttle launch team had completed 92 successful processing flows leading to a launch. The 93rd processing flow would culminate in the launch of the first US assembly flight – designated 2A, or more commonly, STS-88. The facilities and

TABLE 4: *ISS Shuttle Missions Processing Flow.*

STS	OV	OPF	VAB	Pad	Roll back	Launch
88	105	1998 01 Feb	1998 15 Oct	1998 21Oct	-	1998 Dec 4
96	103	1998 07 Nov	1999 15 Apr	1999 23 Apr	1999 16 May	-
			1999 16 May	1999 20 May	-	1999 27 May
101	104	1998 27 Sep	1998 Dec 10 (storage)	-	-	-
		1999 17 Feb	1999 Aug 25 (storage)	-	-	-
		1999 Sep 24	2000 17 Mar	2000 25 Mar	-	2000 19 May
106	104	2000 29 May	2000 07 Aug	2000 13 Aug	-	2000 08 Sep
92	103	1999 27 Dec	2000 21 Aug	2000 11 Sep	-	2000 11 Oct
97	105	2000 24 Feb	2000 25 Oct	2000 31 Oct	-	2000 30 Nov
98	104	2000 19 Sep	2000 04 Dec	2000 02 Jan	2000 02 Jan	-
			2000 02 Jan	2000 03 Jan	2001 19 Jan	-
			2001 19 Jan	2001 26 Jan	-	2001 07 Feb
102	103	2000 24 Oct	2001 02 Jan	2001 12 Feb	-	2001 08 Mar
100	105	2000 12 Dec	2001 19 Mar	2001 22 Mar	-	2001 19 Apr
104	104	2001 04 Mar	2001 29 May	2001 20 Jun	2001 20 Jun	-
			2001 20 Jun	2001 21 Jun	-	2001 12 Jul
105	103	2001 21 Mar	2001 13 Jun	2001 01 Jul	-	2001 10 Aug
108	105	2001 09 May	2001 24 Oct	2001 30 Oct	-	2001 05 Dec

experience of preparing the Shuttle for orbital flight were now well tried and tested and had recently seen a remarkable reduction in launch delays and aborts since the 1988 return to flight mission of STS-26 following the 1986 Challenger accident. But in the first 18 months of Shuttle operations to ISS, awaiting the launch of Russian elements was only one cause of delay. Elements of the US hardware were also slipping in the processing schedule and in the summer of 1999, the problems in the Shuttle wiring subsystem also affected the processing flow for ISS. While these difficulties certainly delayed the permanent habitation of ISS, once the Russian Service Module Zvezda had been launched and the wiring problem had been resolved, a flight rate of four Shuttle missions in the first 18 months of ISS orbital operations (Nov 1998-May 2000) was followed by a flow rate of 9 missions in the next 18 months (through Dec 2001).

Of course, the complexity of the systems involved meant that the ISS Shuttle ground processing didn't all go without a hitch. On STS-88, the initial launch attempt on 3 December 1998 was aborted after a sudden drop in pressure in one of three hydraulic systems on the orbiter, when the APUs shifted from low to high pressure in the hydraulic lines after start up. With the engineers unable to source the problem within the launch window, the launch was abandoned for 24 hours. On 8 May 1999, a hailstorm damaged the foam insulation on the ET of STS-96 and

Canada

Workers in the Orbiter Processing Facility monitor the lowering of the Keel Yoke Device (KYD) used to support the International Cargo Carrier into the payload bay of the orbiter. To the right of frame is the orbiter docking facility (with a red temporary launch processing protective cover) installed in the forward end of the bay and along the far sill of the payload bay lies the Canadian remote manipulator. Processing support platforms surround the payload bay area and note the protective clothing and restrain harness to prevent damage to flight equipment by workers.
(NASA)

closer inspection revealed that repairs could not be performed on the Pad, so the stack was returned to the VAB and the launch rescheduled. A total of 648 divots were recorded, with 211 needing repair, 248 receiving patches and 189 deemed safe for flight. The launch (planned for May 20) was delayed seven days.

When STS-93 was launched in July 1999, a voltage drop in Columbia's electrical system just 5 seconds after launch was compounded with a hydrogen leak in Engine No. 3, almost resulting in a tricky Return to Launch Site Abort procedure. The resulting inspections revealed damage to orbiter wiring across the fleet, due to loss of insulation as a result of repairs in ground processing over 20 years. The inspection of the faulty engine revealed a severed minute post pin plug, used to seal a LH leak in a supply loop. This struck and damaged the engine LH supply system. Post-flight repair and prevention programmes would take the rest of 1999, so STS-101 was slipped to 2000 [4].

The new launch for STS-101 was set for the spring of 2000, but delays with the Russian Service Module forced NASA to re-examine the mission manifest and the possibility of splitting the mission before finally inserting STS-106 into the launch schedule after the launch of Zvezda. Ironically, the launch of STS-101 was delayed an extra week due to sprained ankle sustained by Commander Halsell on 15 March. Then bad weather forced scrubs on three successive days (24, 25 and 26 April) before a further reassignment from 16 to 19 May due to the launch scrub of an unmanned Atlas III/EUTELSAT launch and a NASA/USAF agreement on launch priorities.

The launch of STS-106 was threatened by 'Hurricane Conditions' during August 2000, primarily from Hurricane Debby, but the storm was down graded to a tropical wave with only lightning threatening. Though the lightning mast on Pad 39B did record a strike on 5 September, but no damage was reported and this had no impact on the launch.

During STS-92 preparations, the planned launch on 5 October 2000 was delayed due to the potential problem from an exploding bolt, observed during the separation of the ET on 106. It was determined to postpone the launch 24 hours to 6 October to review the situation. During this hold, a performance problem was noted with a valve in the MPS, which had to be removed and replaced. The launch now slipped to 9 October, but high winds prevented loading of the ET and delayed this a further 24 hours. The rescheduled launch on 10 October was also scrubbed when a pin used to secure handrails at the launch pad was found to be lodged on a strut connecting Discovery to the ET, forcing yet another 24 hour delay.

The majority of STS-97 processing proceeded normally, but with STS-98 new problems surfaced to threaten the 2001 launch schedule.

During the Christmas holidays of 2000, an extensive evaluation of SRB ordinance cables revealed minor damage. This was quickly repaired and cleared for flight. Then, only one hour after leaving the VAB on 2 January 2001, the STS-98 stack was halted on the crawler way to allow inspection of a failed computer processor on the number 1 crawler transporter. The repair could not be completed where the vehicle stopped, so the stack was rolled back to the VAB and exchanged to Crawler Transporter no 2. The stack was then successfully taken to the pad the next day. However, while processing of Atlantis continued on the pad, additional testing and analysis of SRB cables had continued across the fleet. When uncertainly over the integrity of cables assigned to STS-98 was expressed, a decision to delay the launch into February resulted in a roll back to the VAB for closer inspections. The Destiny Laboratory remained at the Payload Changeout Room at Pad 39A while the stack was returned to the VAB. The inspection conducted in the VAB involved X-ray imaging and so called 'wiggle tests', where cables were shaken by hand while still in the umbilical tunnels!

With no visible physical damage, the system was cleared for launch by the Mission Management Team (MMT) on 7 February, a date that offered a better opportunity for ISS rendezvous, but also delayed the launch of STS-102 from 1 March to 8 March. Only minor processing problems were recorded for STS-102 and STS-100, but during STS-104 preparations, three constraints affected the launch processing.

Firstly, the MMT announced on 25 January 2001 that launch would occur no earlier than 8 June. This was beyond the so-called 'beta angle cut-out' of the Sun that would have produced undesirable thermal conditions on the orbiter if in orbit between 18 May and 7 June. By 31 May, problems encountered with the new Space Station Remote Manipulator System (SSRMS) that had been delivered on STS-100 resulted in a further delay of the 104 launch into July to allow time for on-orbit evaluation of the faulty arm. After deciding on 12 July as a launch date, the 104 stack was being rolled to the pad on 20 June when weather reports indicated a threat of lightning in the vicinity of LC39, so the stack was returned to the VAB for the night before finally being taken to the pad the next day.

STS-105 processing was also delayed due to the ongoing robotic arm evaluations during June 2001, but when the stack was finally rolled to Pad 39B on 1 July, processing towards launch progressed smoothly. The final launch of this period, STS-108, also progressed without major incident towards the planned launch date.

Launch Phase

Processing towards a launch had to take into account the flow of both the payload and hardware and also the training and preparation of the flight crew and mission support teams across the globe. The conditions required for launch include adequate launch and landing lighting times, and the availability of contingency landing facilities at the Cape, across the Atlantic at TAL sites and at Edwards AFB and White Sands, and the 60+ contingency landing site across the globe.

TABLE 5: *Shuttle Main Propulsion System Data.*

STS	SSME1	SSME2	SSME3	ET	SRB Set
88	2043	2044	2045	SLWT-097	BI-095PF
96	2047	2051	2049	SLWT-100	BI-100PF
101	2043	2054	2049	SLWT-102	BI-101PF
106	2052	2044	2047	SLWT-103A	BI-102PF
92	2045	2053	2048	SLWT-104A	BI-104PF
97	2043	2054	2049	SLWT-105A	BI-103PF
98	2052	2044	2047	SLWT-106A	BI-105PF
102	2048	2053	2045	SLWT-107	BI-106PF
100	2054	2043	2049	SLWT-108A	BI-107PF
104	2056	2051	2047	SLWT-109A	BI-108PF
105	2052	2044	2045	SLWT-100A	BI-109PF
108	2049	2043	2050	SLWT-111	BI-110PF

Availability of the orbiter fleet is determined by pre-programmed orbiter maintenance and refurbishments phases that nominally last 14 months and effectively take an OV out of the processing flow. The extensive programme of modification is conducted under the Orbiter Maintenance Down Period (OMDP) at the Rockwell/ Boeing facility at Palmdale, and also includes the Orbiter Major Modifications programme. Other annual events that affect Shuttle launch operations are national holidays (with reductions in KSC workforces), and the Leonids meteor shower event that peaks around the third week of November each year.

The decision to incorporate Russia into the Space Station programme meant that the station would fly at 51.6 degrees inclination (the same as that of Mir) instead of 28.8 degrees (planned for the cancelled Freedom). This change suited launches from the Baikonur Cosmodrome in Kazakhstan rather than from KSC in Florida, and had a payload penalty on Shuttle launches from Florida. As a result, a lighter but structurally stronger aluminium-lithium ET was developed to compensate for the loss of payload capacity. From June 1998, the Shuttles used a super lightweight tank that was 3400 kg lighter than the earlier standard ET. In addition, improvements to three of the four Shuttles equipped them to carry ISS hardware to orbit. Atlantis, Discovery and Endeavour would be the ISS assembly fleet, while Columbia was manifested for non-ISS missions.

The initial Shuttle missions from 1981 followed a 'standard insertion', where two OMS burns were initiated to achieve the required orbit. The first, shortly after MECO, raised the maximum altitude, while the second raised the low point of perigee and circularised the orbit. After ten years of flight operations, as performance of the SSME (up to 104%) became better understood, the OMS-1 burn was omitted, with flights using a 'direct insertion' trajectory employing OMS-2 only. The standard insertion is now only flown when a low orbit altitude mission would result in an ET impact point outside the safe limits of direct insertion profiles. The ET impact point from ISS missions is in the south or mid Pacific Ocean, instead of the Indian Ocean for 28.8 degree missions. Using a 35 degree launch azimuth (direction) from KSC allows the upper safety limit of 57 degrees inclination while still flying a safe distance from inhabited land masses during ascent, yet still allows contingency and abort profiles to be safely observed. The payload penalty of changing from 28.8 to 51.6 degrees equates to about 500 lb per degree, or approximately 11,500 lb less payload mass. It should also be noted that any crewmember additional to the standard 5 person Shuttle orbiter crew also affects payload weight by approximately 500 lbs per person, due to the additional life support and crew escape equipment required.

Launch windows for Shuttle ISS assembly missions have nominally lasted 5-10 minutes. The final moments of the Shuttle countdown occur as the station passes over the Atlantic coast of America and the Shuttle enters orbit behind the ISS to begin the long chase towards the station from the lower orbit.

The flight profile of a Shuttle launch follows the now familiar ascent sequence that has been used since STS-1 in April 1981. However, since STS-87 and following the decommissioning of the S-band facilities at the Bermuda tracking station, the vehicle now returns from the Roll-To Heads Down (RTHD) attitude – flown during the first stage (SRB firing) of the ascent to reduce excessive wing loads – to the Roll To Heads Up (RTHU) attitude to communicate with the TDRS network of satellites in the latter stages of ascent.

Shuttle flight operations in the ISS programme got off to an excellent start with the almost flawless launch of STS-88. STS-96 and STS-101 followed this trend, with no major anomalies reported during ascent. One new feature of STS-101 launch operations was the introduction by USA of a one-man submarine to assist in automated booster retrieval and recovery, instead of the previous use of a diving team. The next eight missions also followed nominal ascent profiles to orbit, with only minor infringements of the final countdown being recorded.

The final mission covered in this paper, STS-108, had been planned for launch on 29 November, but difficulties in docking a Progress vehicle to ISS had slipped the launch to 4 December. Weather concerns delayed the launch a further 24 hours. This was also the first Shuttle mission after the 11 September terrorist attacks on America and increased security measures at the Cape were in force, including heightened restrictions to visitors and press coverage, with added DoD support across the KSC area in case of terrorist threats to the launch. In the event, the launch occurred without incident. All 12 missions to ISS in this period were launched successfully.

Rendezvous and Docking

The skill of rendezvous, proximity operations (Prox Ops) and docking with another orbital vehicle was first demonstrated by the Americans during the Gemini-Agena programme (1965-1966) and put to good effect in the Apollo and Skylab programmes (1969-1973). The Russians demonstrated their skills at rendezvous and docking with early Soyuz and Salyut space stations (1968-1977) and in 1975, the US and USSR cooperated in the joint ASTP international docking mission, with a view to following up the mission with a Shuttle docking to a Salyut station [5].

Following ASTP, the Shuttle was developed with the advertised capability to dock with a second vehicle, but without an integrated docking system to achieve it. During 1984-1993, several Shuttle missions demonstrated

This view out of the aft flight deck windows shortly after beginning orbital operations reveals the full cargo bay with the docking facility lower left, Spacehab module at rear and RMS at right.   (NASA)

rendezvous and Prox Ops capability with various satellites and the Hubble Space Telescope, and used the RMS to 'capture' free-flying payloads without ever conducting a hard docking. Meanwhile, the Russians clearly demonstrated their ability to perfect space station operations with the rendezvous and docking of dozens of Soyuz and Progress spacecraft to second generation Salyut and Mir stations (1977-2000). This also included automated docking by unmanned re-supply craft, a skill the Russians first demonstrated back in 1967. However, it was not until the Shuttle-Mir programme (1995-1998) that the Americans finally had the capability to dock a Shuttle with a space station, using the Orbiter Docking Mechanism. These techniques would be put to greater use during ISS and it's worth recalling that 6 NASA astronauts (Bloomfield, Gorie, Halsell, Jett, Wetherbee and Wilcutt) who flew early docking approaches to ISS had also experienced docking approaches with Mir.

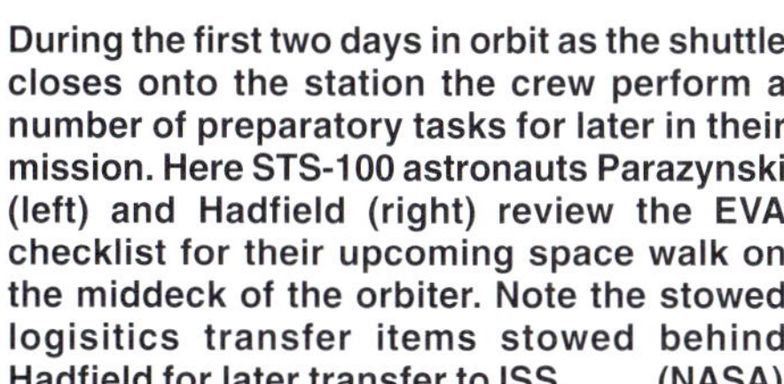

During the first two days in orbit as the shuttle closes onto the station the crew perform a number of preparatory tasks for later in their mission. Here STS-100 astronauts Parazynski (left) and Hadfield (right) review the EVA checklist for their upcoming space walk on the middeck of the orbiter. Note the stowed logisitics transfer items stowed behind Hadfield for later transfer to ISS.   (NASA)

Once the Shuttle has been confirmed for orbital operations, the crew prepare to close in on the ISS over a period of one to two days. Orbiting at a lower altitude than the ISS, the Shuttle uses a combination of thruster burns and orbital mechanics to catch up and match the orbit of ISS by 370 km on each 90-minute orbit. The precise number of burns depends upon the position and time the vehicle enters orbit in relation to ISS, the ascent performance of the main propulsion system (SSME/ET/SRB/OMS), and the orbital position of ISS.

Normally the first and second flight days (FD) are taken up with configuring the Shuttle for orbital flight, testing the RMS, preparing the suits for EVA and lowering the cabin atmosphere from 14.7 psi to 10.2 psi to allow the EVA crew to spend less time pre-breathing pure oxygen (in order to prevent the bends whilst operating in the 4.3 psi environment of their pressure garments).

The Orbiter rendezvous radar is also deployed early in the mission but is not immediately activated. A number of small rendezvous manoeuvres (termed NC burns) can be performed during the final approach, with the crew using the now activated onboard rendezvous radar system to track the ISS, providing distance and closing rates to the crew.

The Commander controls the vehicle from the aft flight deck station as he has the view out of the aft windows into the payload bay (and hence the docking facility), the overhead windows towards the ISS, and from TV cameras mounted inside the docking ring to align the two systems correctly. Assisted by the Pilot (with the rest of the crew ready operate the Shuttle Docking Mechanism and rendezvous ranging devices and an array of cameras), the Commander begins the final approach from about 8 nautical miles directly behind the station. The Terminal Phase Initial Burn (TI) is fired to place the Shuttle on an intersecting flight path with ISS on the next orbit.

Docking approaches:

R-Bar: This is the profile used where the active spacecraft (in this case the Shuttle) approaches its passive target (ISS) along an imaginary line (bar) aligned with the Earth's radius (R). Approaching from 'above' is called 'negative R-bar'.

For Shuttle-Mir, the approach had been from 'below' the station, called 'positive R-bar'. This is more beneficial in docking large objects because in final approach, the differential gravity resulting from the vertical separation acts as a brake and slows the rate of closure. As a safety procedure, and dependent upon the rate of closure, should the active vehicle suffer loss of control of the forward (braking) thruster system, it will be drawn to a halt and then pulled back towards the Earth away from the target, preventing a collision. As with Shuttle-Mir dockings, the 100 tonne mass of the Shuttle and the increasing mass of ISS meant that 'positive R-bar' wa the preferred option during early assembly missions. The help of 'Mother Nature' means the natural tendency for the orbital mechanics is to move the Shuttle away from ISS, thus requiring only a few braking manoeuvres. The R-bar approach from the 'Earth side' also protects direct communications with Russian ground stations.

A view through the orbiter crew optical alignment system (COAS) reveals the gradients used during approaching the station for docking at one of the PMA located visible on the Unity node. (NASA)

The aft flight deck view moments before docking with the ISS. The docking port is visible through the overhead window. (NASA)

V-Bar: This type of approach is in the line (bar) of the orbital velocity vector (V). Here, the active spacecraft (Shuttle) positions itself directly in front of the approaching passive target (ISS) and initiates manoeuvres to approach the target for docking.

For STS-96, the first mission to actually 'dock' to the station, the controllers turned the ISS perpendicular to the Earth so that Unity pointed out to space and Zarya towards Earth. Commander Rominger then brought the Shuttle 600 feet directly 'below' the station in the R-bar mode. The commander then began a half circle, moving the orbiter to 350 feet 'in front' of the station in the V-bar profile and 250 feet directly 'above' the station. At this point, the commander began the descent towards the station until he reached about 170 feet, where the vehicle was held at station keeping until the combination passed within range of Russian flight control communications. After confirmation of the positioning and communication, the orbiter was moved to within 30 feet of the required ISS docking port (on the US element) for a final check of station and Shuttle systems and alignment. The solar arrays on the ISS were normally 'feathered' and locked at the most suitable angle to reduce induced loads at docking, preventing undue damage but also preventing them from gathering solar power. With ISS switched to free-drift at capture and the OV also placed in free drift, this decreased any excess loading on the orbital docking system at point of contact. Using ranging data from instruments and other crewmembers, and a view from a centreline TV camera in the Orbiter docking system relayed to a screen in the Aft Flight Deck, the Commander aligned the cross hairs on the screen with the target on the station docking port. The final closing in speed of one tenth of a foot per second was achieved with an angular alignment of 4 degrees in each axis and a 4 inch lateral alignment, resulting in an initial soft docking and then a hard docking. All connections, seals and pressure checks were completed before the hatches were opened to the mating adapter.

Plume Control: In any approach to the station, slowing down the Shuttle requires firing the RCS towards the ISS creating a plume of exhaust gases that could potentially damage structures (such as solar arrays and antennas), impart a force to make a slight but significant shift in ISS attitude and deposit a layer of residue on the station's surfaces. Therefore, to complete the approach, instead of using the RCS that fire directly at ISS (termed Norm Z firing), jets that are slightly offset in the nose and tail of the orbiter are used (termed Low-Z firing). Low Z is used from about 1000 feet to 250 feet out. From 250 feet, the crew revert to normal operations to have better control of the thrusters.

SNIP or SNOOPY: Two possible approaches (R-bar) employed for docking are when the Shuttle's nose is in the plane with the velocity vector (called Shuttle Nose In Plane, or SNIP) and when the nose of the orbiter is yawed out of plane (Shuttle Nose Out Of Plane, or SNOOPY). As with Mir, this is dependent upon the power levels aboard ISS and the Sun angle at the time of docking.

The Corridor: This imaginary cone begins at 250 feet range, reducing the closer the Shuttle approaches the nominated docking adapter. The reduction in the cone limits the area where Shuttle jets are allowed to fire close to the station and assists in achieving tight docking contact conditions. From 250 feet to 30 feet range, it is a 16 degree cone (35 feet tolerance at 250 feet); from 30 feet to 10 feet range this reduces to a 10 degree cone (2.5 feet tolerance at 30 feet); and from 3 feet to contact it becomes a 6" cylinder.

Following the installation of Destiny on STS-98, docking with the station followed the V-bar approach, with the station parallel to Earth and Destiny to the fore of the flight path with Zvezda to the rear. In this attitude, the Commander moves only a quarter circle around the station until he reaches about 320 feet in front of the PMA attached to the front of Destiny, and then follows the cone approach as before.

The Orbiter Docking System was located in the forward end of the orbiter payload bay, connected to the Spacelab tunnel/Airlock hardware. The physical mating system between the Shuttle and ISS Pressurised Mating Adapters was provided by an Androgynous Peripheral Attach System (APAS), which was a structural ring, a movable ring, alignment guides, hooks, latches dampeners and fixers. It could be mated with an exact copy of itself on the ISS PMA. The Shuttle system was the active element, while those on ISS remained passive. During mating, the Shuttle active element capture ring was extended outwards from the structural ring to capture the passive side of the system on ISS. Any relative motion between the Shuttle and ISS is dampened by attenuation mechanisms. Once all motion is cancelled, the system aligns the two vehicles and then retracts the capture ring (with the passive ring attached) into the structural ring. This activates 24 structural latches to ensure a pressure tight seal for the duration of the docking period.

Docking Operations: Although the STS-88 crew brought together the first elements of ISS and flew a docking system, it did not actually 'dock' with ISS. The RMS was used to help bring the Zarya to the PMA in the cargo bay of the orbiter, which had been attached to the ODS on FD3 (see below). The first true 'docking' with ISS was achieved by STS-96 in May 1999 using the negative R-Bar approach, with Zarya pointed towards Earth and the Unity PMA out towards space.

On STS-101, a positive R-bar approach was completed. Following the recorded 'bad air' situation inside Zarya on STS-96, the 101 crew briefly opened the hatches to take air samples from the mating adapter shortly after docking to confirm the condition of the environment onboard ISS before proceeding inside. STS-106 and STS- 92 followed the positive R-bar approach, while STS-97, also using a positive R-bar, became the first orbiter to dock with the PMA-3 port that had been attached to the Unity module by the STS-92 crew. STS-98 also docked with the PMA 3 module and used the positive R-bar mode.

The V-bar docking of STS-102 was delayed for more than an hour due to one of the feathered solar arrays of the P6 truss failing to indicate a latched signal, a docking requirement to prevent damage to the arrays. Commander Wetherbee held Discovery at the 400 feet point before receiving the go ahead for docking once the ground confirmed the latching and correct lighting conditions to dock. STS-102 docked with PMA-2, now attached to the forward end of the Destiny Module. The next four Shuttles also docked at this port on Destiny in 2001 (STS-100, 104, 105 and 108), all employing positive V-bar approaches.

TABLE 6: *STS Docking Times and Durations.*

STS Mission	Shuttle Orbiter	Date docked	Time docked (EST)	Docking location	Date undocked	Time undocked (EST)	Duration DD: HH:MM	Total hatch open time in ISS DD: HH:MM
88	Endeavour	1998 05 Dec	19:48	Zarya ram*	1998 13 Dec	1525	06:20:38	01:04:32
96	Discovery	1999 29 May	00:24	PMA-2/Unity ram	1999 03 Jun	1839	05:18:17	03:07:30
101	Atlantis	2000 20 May	00:32	PMA-2/Unity ram	2000 26 May	1903	05:18:32	03:08:01
106	Atlantis	2000 10 Sep	01:52	PMA-2/Unity ram	2000 17 Sep	2344	07:21:54	05:09:21
92	Discovery	2000 13 Oct	01:52	PMA-2/Unity ram	2000 20 Oct	1108	06:21:23	01:03:04
97	Endeavour	2000 02 Dec	15:00	PMA-3/Unity nadir	2000 09 Dec	1413	06:23:13	01:00:15
98	Atlantis	2001 09 Feb	11:51	PMA-3/Unity nadir	2001 16 Feb	0906	06:21:15	02:15:09
102	Discovery	2001 10 Mar	01:38	PMA-2/Destiny ram	2001 18 Mar	2332	08:21:54	05:22:22
100	Endeavour	2001 21 Apr	09:59	PMA-2/Destiny ram	2001 29 Apr	1334	08:03:35	Data not available
104	Atlantis	2001 13 Jul	23:08	PMA-2/Destiny ram	2001 22 Jul	0054	08:04:46	Data not available
105	Discovery	2001 12 Aug	14:42	PMA-2/Destiny ram	2001 20 Aug	1052	07:20:10	Data not available
108	Endeavour	2001 07 Dec	15:03	PMA-2/Destiny ram	2001 15 Dec	1228	07:21:25	Data not available

Notes: * STS-88 unberthed the Unity module from the payload bay and attached it to the Shuttle docking ring, and then latched on to Zarya with assistance of the RMS. EST - Eastern Standard time.

ISS-2 crew member James Voss (centre) congratulates STS-102 command Jim Wetherbee, at the aft flight station, on a successful docking with ISS, and in safely delivering him to his new home for the next few months. (NASA)

Payload Delivery

In addition to the major payloads delivered to ISS, each Shuttle carried a range of secondary and supplementary payloads and experiments that were not always part of the mainline ISS science programme.

The major secondary payloads carried on the Shuttle ISS missions (see also Andy Salmon's paper on ISS science investigations on page 119-143) included:

STS-88: IMAX Cargo Bay Camera (ICBC); Satellite de Aplicaciones/Cientifico A (SAC-A), an Argentinean non-recoverable satellite; Mighty Sat 1 non-recoverable deployable (using Hitchhiker support equipment); and the Space Experiment Module (SEM) [SEM-07], an educational initiative to increase student access to space from Kindergarten to University levels.

STS-96: Integrated Vehicle Health Monitoring HEADS Technology Demonstration 2, to evaluate the health of shuttle systems in flight; Shuttle Vibration Forces Experiment that measured dynamic forces between the orbiter and canisters held against the payload bay wall; and Student Tracked Atmospheric Research Satellite for Heuristic International Networking Equipment (STARSHINE), one of a series of satellites to study the effects of the Sun's cycle on Earth's atmosphere.

TABLE 7: *Launch/Orbiter/ Payload Weights (lbs) (Prelaunch figures from NASA Press Kits).*

STS Mission	Shuttle Lift-off	Orbiter/Payload Lift-off	Orbiter/Payload landing	Orbiter delivered Primary Payload(s)
88 (1A)	4,518,390	236,927	200,296	Unity Node 1/PMA 1/PMA 2
96 (2A)	4,514,454	262,035	220,980	Spacehab DM; ICC
101 (2A.2a)	4,519,492	262,565	224,504	Spacehab DM; ICC
106 (2A.2b)	4,519,645	254,099	221,271	Spacehab DM; ICC
92 (3A)	4.520,596	253,807	204,455	Z-1 Truss; CMGs; Ku/S-Band; PMA3
97 (4A)	4,524,410	266,185	197,879	PV Module P6
98 (5A)	4,520,235	254,694	198,909	Destiny US Laboratory
102 (5A.1)	4,522,922	219,363	198,507	MPLM-1 Leonardo (01)
100 (6A)	4,517,989	257,748	222,275	MPLM-2 Raffaello (01); Canadarm2
104 (7A)	4,520,042	258,222	207,251	Quest Airlock
105 (7A.1)	4,521,931	228,188	219,890	MPLM-1 Leonardo (02)
108 (UF-1)	[data not available at time of publication]			MPLM-2 Raffaello (02)

STS-101: Bio Tube Precursor Experiment to test new technologies planned for the STS-107 research mission; Mission to America's Remarkable Schools, a life sciences payload containing 20 experiments from schools across the US; SEM 6, containing ten passive experiments from schools as part of an educational initiative programme using a GAS canister.

STS-106: Commercial Generic Bio processing apparatus designed to investigate how biological processes are affected by microgravity; SEM 8 carrying 13 school experiments.

STS-92: IMAX Cargo Bay 3D Camera.

STS-98: SIMPLEX, using ground based radar to observe ionospheric disturbances created by OMS thrusters firing.

STS-102: Passive Dosimeter System; Space Experiment Module Carrier Systems; Wide Band Shuttle Vibration Force Measurements.

STS-104: IMAX Cargo Bay 3D Camera; EarthKAM student Earth observation programme.

STS-105: The Materials ISS Experiment, the first externally mounted experiment on ISS designed to fly material and space exposure experiments in a low cost non-intrusive mode to test critical components for future applications; Small Payloads Take Youth Science into Orbit, including a Simple Satellite and a microgravity smouldering combustion; and SEM 10 with 11 student experiments.

STS-108: Avian Development Facility – Japanese Quail Eggs; Commercial Biomedical Test Module experiment and Starshine 2.

Several GAS canisters were also flown on the missions. Getaway Special (GAS) is a long series of small experiments that has been flown in canisters in the cargo bay of the Shuttle on various missions since 1982. Those flown on ISS missions were:

STS-88: (G-093R) Vortex Ring Transit Experiment – designed to investigate the process whereby liquid is converted into small droplets.

STS-106: (G-782) An education project to allow elementary and high school students from eight schools in the St. Louis area to have hands on space science experience.

STS-102: (G-783) Also known as Aria-2 is a second educational project containing 124 passive science experiments to encourage students in the St. Louis area to use hands on science experimentation.

STS-105: (G-774) designed to increase the understanding of smouldering combustion in a long-term microgravity environment; (G-780) investigations of crystal growth in microgravity.

Several middeck and Spacehab located experiments were also flown:

STS-101 carried a Biotechnology Ambient Generic experiment designed to grow high quality protein crystals in microgravity; a commercial protein crystal growth experiment to grow crystals of human alpha interferon to be used against several afflictions; and a Gene Transfer Experiment using the astroculture Glove Box to produce commercially important transgenic plant materials in microgravity.

STS-106 re-flew the Middeck Active Control Experiment that provides useful data on reducing the effects of vibration in moving structures in space. This became the first active science payload to fly aboard ISS. The flight also carried the Protein Crystal Growth Enhanced Gaseous Nitrogen Dewar experiment to demonstrate a low cost platform for conducting a number of experiments designed to define the optimum conditions for the growth of large, high quality protein crystals in microgravity.

TABLE 8: *Detailed Test Objectives (DTOs)/ Detailed Supplementary Objectives (DSOs).*

DTO	Objective	Assignments
257	Structural Dynamics Module Validation	STS-88; 97; 102
261	ISS On-orbit Loads Validation	STS-97; 102; 104
262	On-Orbit Bicycle Ergo meter Loads Measurement	STS-104; 108
263	Shuttle Automated Reboots turning	STS-98; 102
623	Cabin Air Monitoring	STS-101
675	Incapacitated EVA Crewmember Translation	STS-98
686	Heat Exchange Unit Evaluation	STS-96
689	USA Simplified Aid for EVA Rescue (SAFER) demonstration	STS-88
690	Urine Collection Device (UCD)	STS-96
692	ISS waste Collector Subsystem Refurbishment	STS-104
700-14	Single String Global Positioning System (GPS)	STS-88; 96; 101; 106; 97; 98; 102; 104; 105; 108
700-15B	Space Integrated GPS/Inertial Navigation System (SIGI)	STS-88; 96
700-21	SIGI Orbital Attitude Readiness	STS-101; 106
700-22	Crew Return Vehicle (CRV)/ SIGI	STS-97; 108
701	Space Vision Laser Camera System	STS-105
805	Crosswind Landing Performance	STS-101; 106; 97; 98; 102; 104; 105; 108
847	Solid State Star Tracker Size Limitation	STS-88; 96; 101
1214	Resource Transfer Line Capability Evaluation	STS-96

Human Exploration & Development of Space (HEDS) Technology Demonstrations – DTO HTD

DTO	Objective	Assignments
1403	Micro-Wireless Instrumentation System (Micro-WIS)	STS-97; 104
1404	Laser Dynamic Ranger Imager (LDRI)	STS-97

DSO	Objective	Assignments
331	Interaction of the Shuttle Launch and Entry Suit (LES) and Sustained Weightlessness on Egress Locomotion	STS-88
490	Bioavailability and Performance Effects of Promethazine during Space Flight	STS-108
493	Monitoring Latex Viral Reactivation and Shedding in Astronauts	STS-101; 106; 102; 104
496	Individual Susceptibility To Post-Spaceflight Orthostatic Intolerance	STS-88; 106; 98; 102; 104
497	Effects of Micro gravity of Cell Medicated Immunity and Reactivation of Latent Viral Infections	STS-88
498	Space Flight and Immune Function	STS-101; 106; 98; 104; 105; 108
499	Eye Movements & Motion Perception Induced by Off-Vertical-Axis Rotation at Small Angles of Tilt After Spaceflight	STS-106
500	Space Flight Induced Reactivation of Latent Epstein-Barr Virus	STS-108
503S	Test of Midrodrine as a Countermeasure Against Postflight Orthostatic Hypotension	STS-108
632	Pharmacokinetics and Contributing Physiological Changes during Space Flight	STS-108
634	Sleep-Wake Autography and Light Exposure During Spaceflight	STS-104
635	Spatial Reorientation Following Spaceflight	STS-104; 105
802	Education Activities	STS-97; 98
904	Assessment of Human Factors	STS-88

STS-97 carried the Jason Onboard Seeds project transferred to ISS during the docked portion of the mission. This used digital photography to document the growth of germinating soybean and corn seeds under various spaceflight conditions.

STS-98 re-flew the BPCG-EGN experiment flown on STS-106 and conducted a series of Earth observations of the Coral Reef, the Lower Nile River in Egypt, the southern oscillation of El Nino, and changes in the Yellow River Delta, part of a long term project covering 1980-2000, the first Crew Earth Observations by photography planned for a large and important programme on ISS resident missions. The Shuttle Ionospheric Modification with Pulsed Local Exhaust (SIMPLEX) experiment used OMS thruster firing to create disturbances in the ionosphere, observed by SIMPEX radars located on the Earth's surface.

Spacehab

Spacehab Inc was formed in 1983 by a commercial consortium that recognised the need to augment the limited volume of the Shuttle middeck with a separate payload module, to support small crew-tended scientific and commercial experiments in space. The first Spacehab single module flew in 1993 and it has since completed 7 separate missions (STS-57, 60, 63, 76, 77 91 and 95). On 23 December 1997, NASA announced the award of a $42.86 million two and a half year contract from JSC to Spacehab Inc., to provide logistical support to ISS missions for three missions, providing integration and operation services (NASA News C97-v).

On each Shuttle flight there are 42 lockers in the middeck, but only 7 or 8 are available for scientific equipment. The other 35 are filled with crew equipment, food, clothing, cameras, tools and so on. With the middeck also serving as the galley, bedroom, bathroom, exercise gym and off duty area as well as a workplace, there was a problem finding room for experimentation and stowage on a mission. During the Shuttle-Mir programme, additional working volume was provided by joining two Spacehab modules together, doubling the available workspace in the Shuttle crew module and increasing the logistical capability to and from the space station. The usefulness of Spacehab as a space station logistics module was clearly demonstrated during Shuttle-Mir and provided the baseline data for planning its use to ISS for the initial logistics period pending the arrival of the first resident crews and the availability of the MPLMs to support them.

TABLE 9: Spacehab Specifications.

Description	Single Module (SM)	Double Module (DM)
Width	13.5 feet	13.5 feet
Height	11.2 feet	11.2 feet
Volume	308 cu ft	1100 cu ft
Launch Weight	10,000 lbs	20,000 lbs
Cargo Capacity	3000 lbs	10,000 lbs
Lockers	61 max	61 max
Rack/Lockers	1 rack/51 lockers	61 lockers 4 rooftop locations
	2 racks/41 lockers	4 racks (2 powered), floor stowage for large, unique items
Windows	None	1
Fabrication	Aluminium with multi layered insulation, fits into 1 quarter of payload bay	
Sub systems	Lighting; ventilation; limited power supply; command and data; fire detection and suppression; vacuum venting.	
Connections	Uses Spacelab tunnel and adapters to connect to middeck and Shuttle Docking/ Airlock facility	

Two astronauts are normally assigned responsibilities for the Spacehab Module, with the module environmental control system designed to support two crewmembers on a continuous basis. At the expense of reduced cabin air retraction capability, additional crewmembers can be accommodated in the module for short periods. The Spacehab module is also used as additional location for in-flight experiments and investigations not being transferred to ISS:

Lockers: These are standard mid-deck lockers used on the Shuttle, measuring 10.6 inches tall by 18.4 inches wide and 21.1 inches deep. Two types of trays are available. The larger measures 9.59 inches tall, 16.95 inches wide and 20.0 inches deep, while the smaller measures just 4.5 inches tall but with the same width and depth as the larger tray. Netting covers the top of each tray to prevent items floating out, but with zip-lock openings to allow single item insertion or removal by a crewmember. Alternatively, lockers have moulded foam inserts to support delicate items such as individual tools and cameras. Lockers can support 60 lbs in their 2 cu feet volume. Power is available at 115 watts, 28 volts DC. Above 60 watts heat rejection must be provided by customer-provided cooling systems. Vacuum venting and data facilities are available on a mission dependent basis.

Mounting Plates: These are available for mounting experiment packages as an alternative to the lockers. They are flat plates, identical to those used in the middeck, and feature two optimal hole patterns for fixings. The plates are categorised as single (60 lb capacity) or double (120 lb capacity), depending upon the number of locker spaces occupied. With similar resources available as the locker-contained experiments, these plates offer standard interfaces while giving maximum flexibility in the design of experiment carriers.

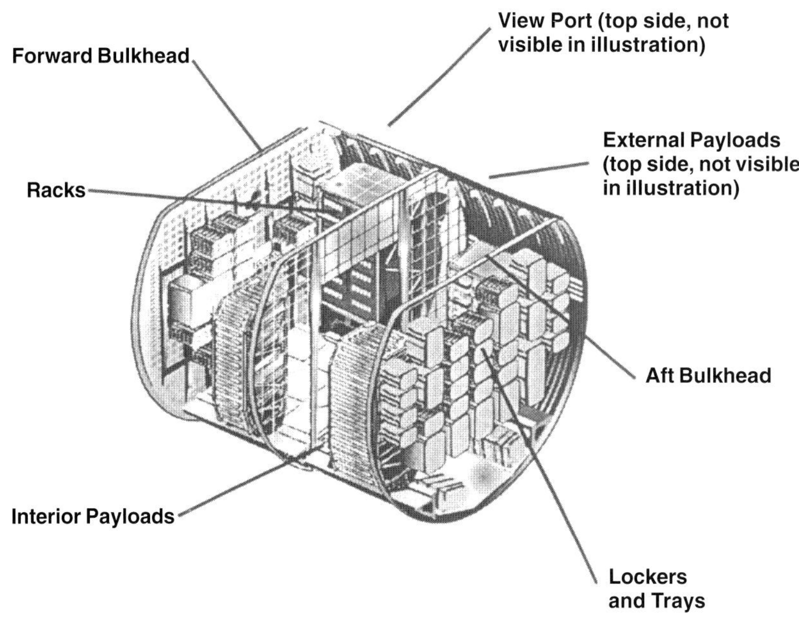

SPACEHAB Logistics Double Module Specs

- 20 feet long, 14 feet wide, 11.2 feet high
- Volume of 308 cubic feet
- STS-106 launch wt. approx. 18,000 pounds
- View Ports: 2
- Fabricated of aluminium multi-layer insulation
- Can accomodate 61 lockers, four rooftop locations
- Can house four racks (two powered)

Integrated Cargo Carrier Specs

- **Eight feet long, approx. 15 feet wide, 10 inches thick**
- **STS-106 launch wt. approx. 4,500 pounds**
- **Fabricated of aluminium**
- **Can accomodate six 400-lb. capacity boxes on a pallet surface**
- **Cargo can be carried on both faces of pallet, top and underside**

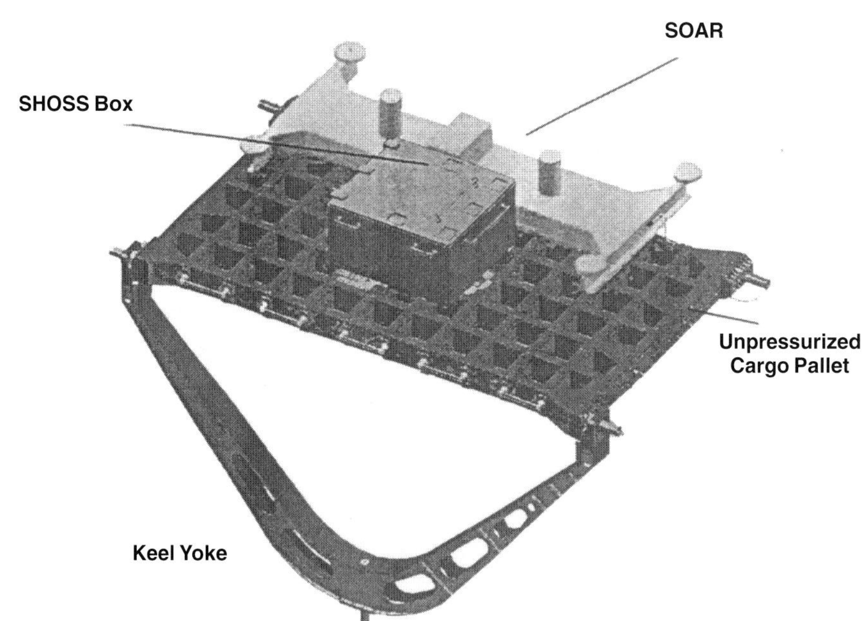

Spacehab Logisitics Double module and Integrated Cargo Carrier (STS-106 specs).   (Spacehab Inc/ Boeing)

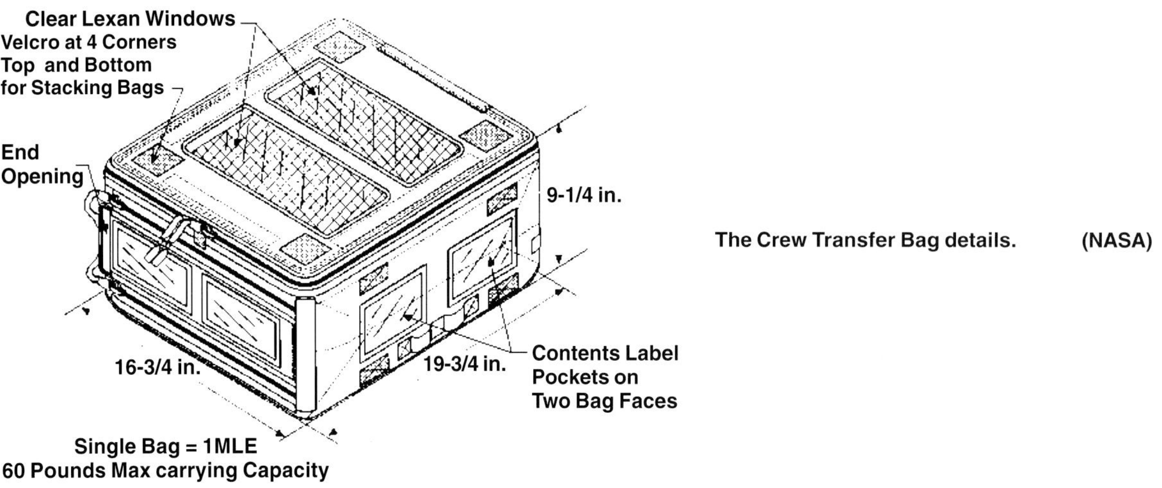

The Crew Transfer Bag details.   (NASA)

Racks: Available in single or double configurations, the single rack measures 80.0 inches tall by 22.1 inches wide by 35.0 inches deep and the double measures 80.1 inches tall by 42.1 inches wide and 35.0 inches deep. The payload accommodation varies from 655 lbs for the single rack up to 1210 lbs for a double rack, with each rack having 1000 watts of DC power available, plus water cooling and vacuum venting. Each incorporates standard 19-inch hole patterns for front panel mounting. The Racks are compatible with either spacelab science modules or Space Station. A single rack replaces the volume of ten lockers inside the Spacehab.

Soft Stowage Bags: Mounted to the plates in front of stowage racks or to the module floor, these contain additional soft transfer items for launch, such as clothing, linen, and items too large for the Crew Transfer Bags and water bags, prior to transfer to ISS.

Crew Transfer Bags (CTB): Available in four sizes; half, single, double and triple, which relates to the Middeck Locker Equivalent (MLE) dimensions.

Type CTB	Capacity Max	Dimensions in Inches
Half	30 lbs	16.75 x 9.75 x 9.25
Single	60 lbs	16.75 x 19.75 x 9.25
Double	120 lbs	18.75 x 19.75 x 18.75
Triple	180 lbs	18.75 x 19.75 x 28.0

Each bag has an end opening facility and clear lexan windows for visual observation of content and also displays a content label in English in pockets on two faces of the bag. The label provides location information (colour coded for quick stowage in various locations in the ISS, e.g. Salmon for Node 1, Tan for FGB) or for return to Earth ('Go Home') at the end of the mission. White labels are reserved for bags packed on orbit, with white cards used by the crew to manually detail what they put inside.

There is a general description of what is in the bag (clothing, tools etc.), a detailed equipment list, and quantity and toxicity levels of the most toxic item in the bag. Every item in the bag is labelled with a bar code identification label, operations nomenclature and toxicity level.

Spacehab Double Modules flew on the logistical supply flights (STS-96, 101 and 106), delivering cargo and supplies to Zarya and Zvezda prior to the arrival of the first resident crew.

Cargo Manifest:

STS-96: A total of 115 items (weighing 3,718lbs) included clothing, sleeping bags, spare parts (including replacing 12 of 18 Russian battery controllers in Zarya), medical equipment, resident crew supplies and hardware. In addition, 18 items (197 lbs) was transferred from ISS back into the Spacehab for the return to Earth. The crew also transferred approximately 84 lbs of water into the station in 7 bags.

STS-101: Delivered 4,900 lbs of Russian and US supplies, including clothing, personal hygiene articles, healthcare supplies, exercise equipment (treadmill and ergometer), food, TV and movie equipment, a fire detection and suppression system, computers and sensors, and replacement storage batteries for Zarya.

STS-106: 8,100 lbs of hardware equipment and logistical supplies completed the outfitting of ISS, including water transfers and 762 lbs of equipment back into Atlantis.

MPLM

Often called the Space Station Moving Van, the Multi-Purpose Logistics Module (MPLM) is the Italian built equipment transfer module carried into orbit in the orbiter payload bay. It is transferred to the ISS Node by the RMS for unloading and then returned to the payload bay at the end of docked activates for the return to Earth. There are three MPLMs available for use on ISS missions. The first, called Leonardo, flew on STS-102 and 105; the second, Raffaello, flew on STS-101 and 108; the third module, Donnatello, was not used during the STS operations covered in this paper.

Each module weighs 4.5 tons, measures 21 feet in length and 125 feet in diameter, and has capacity for 16 laboratory racks. The experiment racks are fitted out with scientific equipment for the resident crew research programmes. The modules can also carry racks and platforms loaded with bags of supplies and logistics to and from the ISS. When racks and science equipment on ISS have completed their programme or need to be replaced they can be removed from the station and returned to Earth for refitting and reuse later in the programme.

There is no access to the module from the crew compartment, and only after the Shuttle has docked to the station will the astronauts use the RMS to transfer the module to the ISS Node for unloading through internal hatches directly into the station. The Module's life support system, fire and suppression sub system, power

The Multi-Purpose Logistics Module (MPLM) called Raffaello is suspended above a workstand in the Space Station Processing Facility (SSPF). (NASA)

Leonardo Multi-Purpose Logistics Module (MPLM) rests in Discovery's payload bay. (NASA)

supplies and computer software can support two astronauts during loading and unloading, as well as any flora or fauna that happens to be part of the science payload carried.

Re-supply and Stowage Platforms are supplied with Cargo Transfer Bags filled with additional equipment and supplies for the station, adding to the overall tonnage lifted to and from the station. Not counted in the CTB launch weight figures are the foam packaging used to retain cargo in the bags, or the straps and soft netting (or fences) used to restrain the bags in position for launch, in orbit and landing.

The lack of access to the Module from the Shuttle crew compartment is advantageous for some experiments and research programmes, where the less variables affecting the experiments in the transit from Earth to ISS, the better. In addition, the route to ISS from the attached module is a more direct one and does not require negotiating the tight passageway from the payload bay through the Shuttle tunnel and docking module and through several hatches into the station. It therefore allows easier passage of larger items, such as the International Standard Payload Racks, which can be fitted out on the ground and handled into the station complete, saving valuable crew time. Each of the Payload Racks measures 73 inches high and 42 inches wide, and contains 53 cubic feet of volume. Often described as 'refrigerator sized', they can carry 1,540 lbs of experiment hardware.

Cargo Manifest:

Leonardo flight 1 (STS-102): Supplies and equipment to outfit Destiny including: six ISS systems racks, three re-supply racks, four re-supply stowage platforms, and the first of the experiment racks.

Raffaello Flight 1 (STS-100): Four re-supply and stowage racks; four supply stowage platforms; two scientific racks (**EX**pediate **PR**ocessing of Experiments to the Space Station, or EXPRESS, Rack 1 and 2).

Leonardo Flight 2 (STS-105): 12 racks, including two EXPRESS racks, six re-supply stowage racks and four re-supply stowage platforms filled with logistics and supplies required by ISS.

Raffaello Flight 2 (STS-108): Filled with equipment and supplies to outfit Destiny, including eight re-supply storage racks and four re-supply stowage platforms.

Integrated Cargo Carrier

The Integrated Cargo Carrier is an unpressurised flat bed pallet located in the Shuttle's payload bay to expand the capability to transport cargo the ISS. Constructed from aluminium, it features a keel-yoke assembly and measures 8 feet in length, 15 feet wide and only 10 inches thick, with a launch weight of approximately 3000 lbs. Capable of supporting up to 6000 lbs of cargo, the ICC was flown on the STS-96, 101, 106 logistics re-supply missions (along with the Spacehab double module described above) and on the STS-102 crew rotation mission.

Cargo Manifest:

STS-96: 3,100 lbs including the Orbital Replacement Unit Transfer Device, the US built crane; elements of the Russian external Strela transfer crane; and the Spacehab Oceaneering Space System Box (SHOSS) – a logistics items carrier.

STS-101: 3,700 lbs including further elements for Strela, the Space Integrated GPS / Inertial Navigation (SIGI); Space Orbital Altitude Reference (SOAR); and a SHOSS filled with EVA tools and logisitics to be transferred and stowed on Unity.

STS-106: 2,835 lbs payload included a SHOSS carrying up to 400 lbs of EVA equipment.

STS-102: Carried the Early Ammonia Servicer; the Rigid Umbilical; the Lab Cradle Assembly (a Spacelab pallet) supporting the Module to Truss Structure Attach System-A; the Pump Flow Control Assembly and External Stowage Platform.

Orbital Operations

During docked operations at ISS, the Shuttle serves as the primary residence for the orbiter flight crew, with them using the facilities onboard the middeck. Some meals are taken with the resident crews during periods of joint activities. The hatches between the two vehicles are not always open the entire time the Shuttle is docked to ISS, because of Shuttle-based EVAs and to limit the drain on Station consumables.

It was not until the delivery of the Russian Service Module that a crew could reside aboard ISS without an orbiter docked to the station. During the first five missions with no resident crew aboard, the Shuttle crew also took on maintenance and housekeeping aboard the station and performed several installation set up and 'get-ahead-tasks' activities in advance of the first resident crew arrival. On STS-106, this included unloading the first Progress (M1-3) to dock to the station.

As well as completing assigned assembly tasks, the Shuttle crew also helps with the transfer of logisitics to and from the station, for both the resident ISS crew and future visiting Shuttle and Taxi crews. They also perform any experiments and research activities assigned to their flight (including DTO and DSO objectives), as well as assisting the resident crew with removing any unwanted trash and redundant equipment back to the Shuttle for the return to Earth.

Steering jets onboard the orbiters are used in short pulse firings as required over several days, to raise and adjust the ISS orbital parameters (See Phil Clark's paper on ISS Orbital Manoeuvres on pp.**149-159**). Shuttle engine burns can also be used to nudge ISS away from intersecting orbits of any identified debris during periods of docked operations. The Shuttle's life support system has also been employed to supplement that of the station, and on the early missions was used to raise the internal air pressure aboard the station, depleted as a result of repeated hatch openings and crew transfers.

During the 6th, 7th and 8th flight days of STS-96, as they transferred items into the Russian Control Module, the

Despite a busy flight plan during each Shuttle mission the crews still have to perform routine maintenance and housekeeping tasks almost every day. (NASA)

Crew conferences are frequency aboard the Shuttle as the crew review their flight plans and checklists and amend to real time situations and development in order to complete their mission effectively and safely. (NASA)

crew experienced the symptoms of a headache, burning or itching eyes, a flushed face, nausea, and in one case, vomiting. Thought to be linked to the degraded air inside the Zarya, they did not immediately report the incident, which they coped with on orbit. They found no problems inside the Shuttle or in Unity, but when two or more astronauts worked in close proximity or in confined spaces in Zarya the symptoms developed, usually at the end of the working day. They found it took 10-15 minutes to recover back in the Shuttle with mild medication from the onboard medical kits, and the problem was also alleviated when fans were used to circulate the air in the Zarya. Air samples were not taken at the time, which did not help in understanding the problem after the flight. On STS-101, the crew took air samples before completely entering the station.

It was determined that opening the panels on the Zarya restricted airflow through the module. In addition, unlike STS-88 whose crew had experienced no such problems, STS-96 flew the first Spacehab module and it was thought that the added habitable volume this created in the Shuttle restricted the amount of air being pumped into the station elements. Since there was no Service Module docked at that time to supplement air supply, the Shuttle/Spacehab combination was insufficient to supply Zarya when the panels were open for long periods of time. To help study the effects of what the Russians called 'normal out gassing of the station modules', a doctor was added to STS-106 (an assignment in fact delayed from STS-88) [6].

The sheer complexity of each shuttle missions and the software it displayed in this revealing view of the additional laptops and personal computers integrated into the shuttle computer network located on the aft flight deck behind the pilot station.   (NASA)

During the first joint flight with a resident crew aboard, the STS-97 crew followed separate sleep patterns in order to match their two separate work cycles ahead of, and then during the docked operations, only 'visiting the neighbours' for a few hours after 6 days of docked operations.

During STS-102, the first flight of the **MPLM**, the mission was extended by a day to provide additional time for docked activities. It gave the crew more time to complete the logisitics transfer and controlled additional time to review the transferred mass and placement of items stowed in the module for the return to Earth, to ensure it fell within the landing guidelines for a loaded orbiter. The Lead Flight Director for the mission, John Shannon, explained that this also reflected the maturity of the programme in the flexibility of station era time lines as well as the availably of sufficient stores of fuel and life support consumables on board and the progress of crew time lines. However, as the Shuttle remained docked longer than planned, and with two of four computers on the flight deck turned off as a normal procedure during docked operations, the orbiter cooling system became too cold and began to ice up. Turning on the computer allowed additional electronics to thaw out the ice, but a precautionary check of flight deck computers was completed the next day to ensure the software was unaffected.

Shuttle crews have also conducted several press conferences with national and international media services and dignitaries during the relatively few quite moments of their missions.

RMS Activities

The Remote Manipulator System has been an integral element of Space Shuttle operations since the second flight in November 1981, and had played an important part in the initial assembly missions for ISS. The checkout of the RMS primary and backup systems and hardware is one of the many tasks performed by the crew as they approach the station in the first couple of days of the mission.

During STS-88, MS Nancy Currie operated the RMS on its first ISS task, lifting the Unity Module out of Endeavour's payload bay on FD3 and mating it with the Orbiter Docking System. The following day, Currie guided the RMS to grapple the Zarya module and align it with the PMA-1 unit on Unity, setting the arm in the 'limp' mode as the Shuttle's thrusters were fired to bring both spacecraft together. Unfortunately, the initial attempt to hard dock with the arm still attached to Zarya was not totally successful and it had to release its grip on the Russian module to allow the hooks and latches to engage and secure a permanent coupling.

The RMS is also used on the Shuttle-based EVAs, supporting both the astronauts and their equipment on the Articulating Portable Foot Restraint as they go about their tasks. The RMS TV system has provided a useful additional mode of observation during EVAs and photo surveys of the station and orbiter. During STS-92 FD4, a short circuit affected the Space Vision System that provided computerised alignment for the arm, and the keel camera had to be bypassed by a crew devised back up system to regain all functions, except that of the keel camera.

Nominally one astronaut is designated primary RMS operator on the Shuttle crew and is backed up by a second or even third crewmember who is equally trained to operate the arm from the aft flight deck station. In the later missions covered in this period, specialisation in moving payloads or assisting EVA operations meant that more than one astronaut was assigned the role of primary RMS operator (see RMS assignments Table 10).

TABLE 10: *RMS/Loadmster Assignments.*

STS	RMS1	RMS2/RMS BUp	Loadmaster/s
88	Currie	Cabana	Krikalev (Zarya)
96	Ochoa	Jernigan	Ochoa
101	Weber	Halsell	Weber
106	Mastracchio	Altman	Burbank
92	Wakata	McArthur	Lopez-Alegria
97	Garneau	Bloomfield	Garneau
98	Ivins	Cockrell	Ivins
102	Thomas	Kelly	Thomas/Voss
100	Hadfield/Parazynski	Ashby/Parazynski	Rominger/Guidoni
104	Kavandi	Lindsey	Kavandi
105	Forrester	Horowitz (EVA)	Barry
108	Godwin (MPLM)	Kelly M. (EVA)	Godwin

During the logistics supply missions of STS 96, 101 and 106, the RMS was used in an EVA support role. On STS-92, the RMS was used to assist in construction of the ISS once the Z1 truss element had been attached to Unity on FD4. On FD6, the RMS was used to move PMA-3 from the payload bay to install it on Unity. During STS-97, the dual training role of RMS operators was demonstrated when the arm was used by Garneau on FD4 to transfer the P6 solar array structure to the Z1 truss. That was secured by the EVA crew and then Bloomfield took over the controls and guided Noriega, on the end of the arm, as he connected cables around the array. On STS-98, Ivins first used the RMS to move PMA-2 from Unity to a temporary location on the Z1 truss, making room to locate the Destiny laboratory module. Destiny was then removed from the payload bay and rotated 180 degrees before gently aligning it for attachment to Unity, where automated bolts tightened and secured the laboratory to the node. On FD6, Ivins used the RMS to reposition PMA-2 from the truss to the end of Destiny, allowing future Shuttle orbiters to dock there.

Thomas used the RMS on STS-102 FD4 to relocate PMA-3 from its Earth facing (nadir) location to the portside berth, freeing the lower port to receive the 'Leonardo' MPLM (which was moved from the payload bay to the Unity facility on FD5). With logistics transfers completed, the MPLM was released from Unity and relocated in the Discovery's payload bay on FD11. RMS activities were more complex for the next mission, STS-100. During FD4, the RMS was used to move a Spacelab pallet containing the first elements of the ISS robot arm (SSRMS or 'Canadarm2') to latch it onto a cradle located on the Destiny lab for unloading. The next day (FD5), the Shuttle RMS was used to attach MPLM Raffaello to the Unity node, where it remained for the next four days before bring returned to Endeavour's payload bay on FD9. FD10 saw cooperative use of robotic manipulators when the SSRMS that had been delivered on the Spacelab pallet was used by ISS crewmember Susan Helms using the control panel in Destiny to move the unloaded pallet back towards the Shuttle payload bay. In the first hand over exercise, Canadian Chris Hadfield on the aft flight deck of Endeavour, then used the Shuttle RMS to grasp the pallet, allowing Helms to release the ISS arm and using the Shuttle RMS to return it to the payload bay.

A second joint manipulator exercise was planned for STS-104 in the installation of the Quest airlock. During FD3, Helms again controlled the ISS arm from the robotic station in Destiny, while Kavandi was at the aft flight deck of Atlantis during a dry run simulator for the next day's activities. Kavandi supported the EVA crew as they configured the airlock for the transfer, and then Helms used Canadarm2 to grapple the airlock and lift it out of the Shuttle payload bay to berth it on the right side of Unity. The SSRMS was used on FD7 to lift the package of high-pressure oxygen and nitrogen tanks from Atlantis's payload bay and hold them in position while the EVA astronaut on the Shuttle RMS fixed them to the airlock. The plan was to move two tanks (one oxygen, one nitrogen) on this date and the other two on FD10, but the move went so well that they moved a third tank (oxygen) as well. The fourth tank (nitrogen) was moved on FD10.

A Canadian 'handshake in space' occurred on April 28, 2001 as the space station robotic arm (right with the Canada logo visible) transferred its launch cradle (a Spacelab Pallet) over to the Shuttle RMS.　　　(NASA)

The complexity of shuttle station joint operations is revealed in this single frame. An astronaut carries equipment while anchored to the end of the Shuttle RMS across the exterior of the Destiny laboratory (right foreground). Meanwhile the Raffaello MPLM is in centre frame and waiting unloading inside the ISS, while the Canadarm 2 robotic manipulator awaits its role in the proceedings at lower left.　　　(NASA)

The Leonardo MPLM returned to space on STS-105 and was moved by RMS to the Unity node on FD4. On FD7, the RMS lifted the Early Ammonia Servicer hardware, mounted on an ICC, up to the P6 truss where the EVA crew installed it. Leonardo was returned to the payload bay on FD10. During the final Shuttle mission covered in this report, the RMS lifted the Raffaello MPLM, on its second flight, out of the payload bay on FD4 to attach it to Unity node. The RMS returned it to the payload bay on FD10.

EVA Activities

The importance of EVA in the construction and maintenance of ISS has long been realised. America began its experience of EVA techniques in Earth orbit during Gemini in 1965-1966, and again on Skylab in 1973/74. Since 1983, Shuttle crews have repeatedly demonstrated EVA skills in the retrieval, repair and servicing of satellites and scientific platforms, such as Solar Max and the Hubble Space Telescope.

Studies conducted in the late 1980s pointed to a massive amount of EVA during the Space Station Freedom programme [7]. Even when the programme was redirected in the early 1990s, EVA was still considered to become one of the major hurdles, especially during the early assembly flights. This expectation of EVA workload continued to be a challenge for training programmes and mission planning and became known as the 'Wall of EVA', with

TABLE 11: *EVA Crew Assignments.*

STS	EV1	EV2	EV3	EV4	IV1	IV2/IV BUp
88	Ross	Newman	-	-	Sturckow	
96	Jernigan	Barry	-	-	Payette	Husband
101	Williams	Voss	(Horowitz served as BUp EV astronaut)		Horowitz	
106	Lu	Malenchenko	(Burbank served as BUp EV astronaut)		Burbank	
92	Chiao	McArthur	Wisoff	Lopez-Alegria	Wisoff	Chiao
97	Tanner	Noriega	-	-	Garneau	
98	Jones	Curbeam	-	-	Polansky	
102	Voss*	Helms*	Thomas	Richards	Richards	Helms
100	Hadfield	Parazynski	-	-	Philips	Lonchankov
104	Gernhardt	Reilly	(Hobaugh served as BUp EV astronaut)		Hobaugh	
105	Barry	Forrester	(Sturckow served as BUp EV astronaut)		Sturckow	
108	Godwin	Tani	(Walz served as BUP STS-EV astronaut)		Gorie	

Notes: * Voss and Helms conducted their EVA while still officially assigned to the STS-102 crew before transfer to ISS as Increment 2 crewmembers.

STS-88 astronauts wear their thermal undergarments as they prepare for their EVA in the middeck of the orbiter. (NASA)

almost every Shuttle flight requiring one or more EVA periods. The first EVAs from the Shuttle in 1984-85 focused more on satellite retrieval and repair than construction, and although STS 61-B demonstrated early assembly techniques, it was apparent that more experience would be necessary before Shuttle crews could begin constructing the station. Therefore, during a lull in EVA requirements, several Shuttle flights between 1991-1997 (STS-37, 49, 54, 64, 63, 69, 72, and 87) practised ISS EVA techniques to gain experience of what would follow [8].

All Shuttle missions include a team of EVA trained astronauts for contingency and unplanned excursions (such as the separation of the RMS; closing of payload bay doors etc.), as well as any dedicated (ISS) EVA tasks. The EVA 'team' includes those who will perform the space walk (EV1, EV2 etc), the Shuttle RMS operator, and a back up; an intra-vehicular (IV) astronaut who assists the crew in donning and doffing the suits and equipment, and/or a space walk coordinator who ensures the crew outside is following the timeline as closely as possible and does not miss assigned tasks. The coordinators also assist in adjusting the timeline as required in consultation with the ground controllers. For the more complex EVAs, a 'backup EVA astronaut' is assigned to provide a redundancy to ensure tasks are completed. On more demanding multiple EVAs, the teams rotate on alternate EVAs, allowing a division of labour and training and suitable rest periods to pace activities across the mission.

Photo/TV Identification of EVA crews had been solved as early as Apollo 13, with the addition of red stripes on the Commander's arms, legs and helmet. This continued into Shuttle, but with teams of EVA astronauts performing alternate EVAs, amendments to the simple red stripes were devised. The variations of suit ID assigned to crews have included: Solid White Suit [No Stripes]; Solid Red Stripes; Vertical Red stripes; and Diagonal Red stripes.

TABLE 12: *STS-EVA Suit Identification Markings.*

STS	EV	Astronaut	Suit ID	EV	Astronaut	Suit ID
88	1	Ross	Solid Red	2	Newman	Solid White
96	1	Jernigan	Solid Red	2	Barry	Solid White
101	1	Voss	Solid White	2	Williams	Solid Red
106	1	Lu	Solid Red	2	Malenchenko	Solid White
92	1	Chiao	Solid Red	2	McArthur	Solid White
	3	Wisoff	Vertical Red	4	Lopez-Alegria	Diagonal Red
97	1	Tanner	Solid Red	2	Noriege	Solid White
98	1	Curbeam	Solid White	2	Jones	Solid Red
102	1	Voss JS	Solid Red	2	Helms	Solid White
	3	Thomas	Vertical Red	4	Richards	Diagonal Red
100	1	Hadfield	Solid Red	2	Parazynski	Solid White
104	1	Gernhardt	Solid Red	2	Reilly	Solid White
105	1	Barry	Solid Red	2	Forrester	Solid White
108	1	Godwin	Solid Red	2	Tani	Solid White

Once outside identification of each astronaut is made easier by identification marks on the suits. Shuttle EVA astronauts have worn this type of suit for 20 years and the beginning of the programme. (NASA)

Each Shuttle mission to ISS included at least one period of EVA Operations at the space station:

STS-88: The initial Shuttle-based EVA involved with ISS operations featured the EVA crew completing all the umbilical connections required to activate Node 1. The second EVA required the crew to install EVA translation aids and tools, the early communication system antennas and the routing of the communications cable from Zarya to the starboard antenna. The third EVA focused on supporting future EVAs on other missions, including the installation of a large tool stowage bag and repositioning of the portable foot restraints (PFR). In addition, the PMS-2 umbilicals were disconnected, allowing for easier relocation in the future. During the first EVA, Ross set a new American EVA record for accumulated time of 29 days 41 minutes, beating the previous record of Tom Akers in his five EVAs on STS-49 and 61. By the end of the third EVA, Ross had accumulated an impressive 44 hours 9 minutes during 7 EVAs on STS 61-B, 37 and 88.

STS-96: This crew completed one EVA, during which they transferred and installed two cranes on the outside of the station, as well as two new PFRs and three bags filled with tools and handrails for future EVA operations. From the ICC they removed the US Orbital Transfer Device (OTD) crane installing it on to PMA1. Their next task was the relocation of the Russian Strela and grapple fixture adapter plate from the ICC to a grapple fixture on to PMA-2. The EVA crew also conducted a documented survey of painted surfaces on Unity and Zarya, inspected one of the two early communications antennae on the node and installed a cover on the trunnion pin of Unity. Associated IV tasks included relocating foot restraints inside PMA-1 and the installation of three tool bags for later EVAs.

STS-101: The astronauts conducted a single EVA to complete the assembly of the Russian Strela crane and to test the integrity of the US crane. They also replaced the faulty EECOM antenna and installed handrails and a camera cable across the outside of the station.

STS-106: This EVA crew completed the integration of the recently docked Zvezda with the rest of the ISS configuration. That included translating equipment to the Zvezda and the installation of a magnetometer boom. In total, nine power, data and communications cables were routed between Zvezda and Zarya during the EVA.

TABLE 13: *STS-EVAs.*

STS/ISS EVA Seq	Date that EVA commenced	STS mission EVA sequence	EVA Crew Members	Duration Hrs: Min	Mission EVA total duration	Accumulative STS-EVA duration/Total number EVAs	
1	1998 Dec 07	88-1	Ross/Newman	07:21			
2	1998 Dec 09	88-2	Ross/Newman	07:02			
3	1998 Dec 12	88-3	Ross/Newman	06:59	21:22	21:22	3
4	1999 May 28	96-1	Jernigan/Barry	07:55	07:55	29:17	4
5	2000 May 21	101-1	Williams/Voss	06:44	06:44	36:01	5
6	2000 Sep 10	106-1	Lu/Malenchenko	06:14	06:14	42:15	6
7	2000 Oct 15	92-1	Chiao/McArthur	06:28			
8	2000 Oct 16	92-2	Wisoff/Lopez-Alegria	07:07			
9	2000 Oct 17	92-3	Chiao/McArthur	06:48			
10	2000 Oct 18	92-4	Wisoff/Lopez-Alegria	06:56	27:19	69:34	10
12	2000 Dec 05	97-2	Tanner-Noriega	06:37			
13	2000 Dec 07	97-3	Tanner-Noriega	05:10	19:20	88:54	13
15	2001 Feb 12	98-2	Jones-Curbeam	06:50			
16	2001 Feb 14	98-3	Jones-Curbeam	05:25	19:43	108:37	16
17	2001 Mar 11	102-1	Voss-Helms*	08:56			
18	2001 Mar 13	102-2	Kelly M. -Thomas A.	06:21	15:17	123:54	18
19	2001 Apr 22	100-1	Hadfield-Parazynski	07:10			
20	2001 Apr 24	100-2	Hadfield-Parazynski	07:40	14:10	138:49	20
21	2001 Jul 14	104-1	Gernhardt-Reilly	05:59			
22	2001 Jul 17	104-2	Gernhardt-Reilly	06:29			
23	2001 Jul 21	104-3	Gernhardt-Reilly	04:02	16:30	155:14	23
24	2001 Aug 16	105-1	Barry-Forrester	06:16			
25	2001 Aug 18	105-2	Barry-Forrester	05:29	11:45	166:59	25
26	2001 Dec 10	108-1	Godwin-Tani	04:12	04:12	171:11	26

Notes: EVA times are for Shuttle orbiter crews only.

STS 92: The first extensive EVAs at ISS were carried out by two teams. During the first EVA, Team 1 relocated the S-Band Antenna Support Assembly to the Z1 truss and removed and stowed the Z1 starboard bulkhead thermal shrouds. The Z1 to Node 1 umbilicals were installed and the Space to Ground Antenna dish was also installed on its boom and then deployed. The port EVA Tool Storage Devices were also transferred from the Spacelab Logistics Pallet to the Truss. During EVA 2 (Team 2), primary and secondary cable bundles were connected between Unity and PMA-3 and latches were opened on the Z1 Truss ready for the attachment of the P6 solar arrays during STS-97.

EVA 3 (Team 1) included configuration work at the Z-1 truss, where DC-to-DC Converter-Unit Heat Pipes were installed, the Z-1 feel pin was moved to a new location on the truss, and preparations for installing the P6 arrays continued with the removal of launch locks. Assembly Power Converter Unit jumpers were removed from PMA-2 to make way for four Z1 cables and a starboard EV tool Storage Device was also transferred from the Spacelab Logistics Pallet on the truss. EV4 (Team 2) saw the removal of the grapple fixture on the truss, deployment of a utility tray, the release of umbilical launch restraints and the cycling and opening of the Z1 Manual Berthing Mechanism latches. The astronauts also completed a test flight of the Simplified Aid For EVA Rescue (SAFER) Unit – a self-rescue backpack manoeuvring device – as part of a safety protocol test requirement; but they remained attached to the Shuttle tethers and were accompanied by other astronauts on the end of the RMS.

STS-97: Work on the Z1 truss continued with the next mission which also delivered the P6 solar arrays. The P6 Integrated Truss Structure was installed to the Z1 Truss during EVA 1 and then the crew prepared both the solar arrays and radiator for deployment. During the next EVA they configured the hardware for the supply of power through the P6 arrays into the station, positioned the S-band for station use, and prepared for the arrival of the Destiny lab on the next mission.

EVA 3 included activation of the solar arrays. Repair work was needed to increase the tension on the array blankets, which had failed to deploy properly. They also completed the deployment of a floating potential probe that measured the plasma fields surrounding the station. The crew also performed a number of 'get-ahead tasks' that were planned for future EVAs. Finally, at the very top of the P6 array, they continued a tradition observed by Earth bound construction workers of 'topping out' construction, installing an image of a green tree, rather than the real one used in ceremonies on Earth.

STS-98: During EVA 1, PMA-2 was moved by the RMS. The crew disconnected the Assembly Power Converter Unit power cables from Destiny before it was installed on the station, and then connected power cables from Unity to the science lab. EVA 2 saw the relocation of PMA-2 by the RMS, as the crew worked to install a pressure control assembly vent, pin covers, EVA aids and the laboratory's power and data grapple fixture, where the SSRMS would be located during STS-100. They also installed the shutter and gearbox for the optically pure window on the US laboratory. During their final EVA, the astronauts connected a spare S-band Antenna Support Assembly to Z1 and attached electrical and data collection cables between PMA-2 and Destiny. Finally, they demonstrated the capability for one astronaut to carry a second disabled astronaut to a safe location.

During the first STS-98 EVA, a coolant line that Curbeam was connecting leaked a small amount of frozen ammonia crystals. The leak was quickly stopped and the crystals vaporised and dissipated, but following safety rules, Jones brushed off his suit while Curbeam 'Sun bathed' for 30 minutes so that solar heat could vaporise any crystals that might have remained attached to his suit. Re-entering the airlock, they completed only a partial repressurisation then vented the airlock to flush out any residual ammonia before fully pressurising the lock for entry into the Shuttle. The orbiter crew wore masks for 20 minutes as the orbiter's life support system purged the air. The whole EVA took 1 hour 39 minutes longer then planned as a result. Their third EVA was the 100th by US astronauts and 60th from a Shuttle, and they marked the event with a few comments as, "A tribute to all those people who have done space walks," and to those who planned and designed the suits for Gemini, Apollo, Skylab and now the Shuttle.

STS-102: On this flight, two members of the second increment crew completed the first EVA before they took over running the station from ISS-1. A delay was needed when a foot restraint was lost and another was retrieved from a stowage location. The two astronauts unplugged PMA-3 from the electrical cables of the Unity node and then installed the Density Lab Cradle Assembly. At nearly 9 hours duration, this was the longest EVA ever carried out from the Shuttle. During the second EVA, members of the Shuttle core crew (EVA Crew 2) transferred the Early Ammonia Service to the P6 Truss, installed an external stowage platform and delivered a replacement Pump Flow Control System.

STS-100: This mission featured tasks associated with installing the first elements of the station's remote manipulator system. The first EVA required the crew to install the necessary cables that would feed electrical power, computer commands and video between the workstation in Destiny and the arm located outside. In order to improve future communications during Shuttle-station EVAs, a UHF communications antenna was installed and deployed. Finally, the crew released the bolt that secured the Candaarm2 for launch, unfolded it and prepared it for control from within the station. During the second EVA, they retrieved the Early Communications Antenna, which was no longer required, and prepared the area for the arrival of the Quest Airlock during STS-104. Eight Cable connections were completed to allow the station arm to move across the station, while four others from the pallet to station from EVA 1 were disconnected. Finally, the crew stowed a Director Current Switching Unit to the stowage platform for possible future use, an important electrical component back up system.

STS-104: During the installation of the Quest airlock the EVA crew assisted the RMS operators in removing installation and protective covers while the airlock was still in the Shuttle payload bay. They also installed the high-pressure gas tank attachment points and thermal covers on the airlock before disconnecting a power cable. While the unit was transferred to the station, both EVA astronauts remained inside the Shuttle airlock. When Quest was ready for instillation, they went back outside to assist in the alignment and attachment and then attached power cables to power heaters when it was secured on the station. During EVA 2, three High Pressure Gas Tanks (2 oxygen and 1 nitrogen) were installed onto the airlock. The crew released the attachments and the Canadarm2 took each tank to the station, while the EVA crew, on top of the RMS foot restraints at the airlock location, manually installed each tank as it arrived from the payload bay. Connections between the tanks and the airlock were completed as insulation covers were placed over several airlock features. Their third EVA was the first out from the Quest airlock, to install the final nitrogen tank and attach communication cables for Russian-suited EVAs from the airlock. They also installed handholds and thermal covers on the tanks and grapple fixtures. This EVA also during 32nd anniversary of the famous first lunar landing and surface EVA on Apollo 11.

STS-105: The first EVAs featured the installation of the Early Ammonia Servicer (EAS) onto the P6 Truss. They also installed the dual containers of the Materials ISS Experiment (MISSE) on the exterior of Quest, which will be retrieved on a later EVA. The second EVA involved installing handrails onto Destiny and heater cables were also attached to the laboratory in preparation for use on STS-110 in 2002, during which power heaters are to be installed on the station's S0 Truss segment.

STS-108: The final EVA by Shuttle crews during this period was during this mission, where insulation blankets were installed around two cylindrical Beta Gimbal Assemblies – used to rotate the large arrays – and several 'get-ahead-tasks' were completed in preparation for STS-110. The astronauts could only be moved part the way up the truss by the Shuttle RMS before having to revert to hand over hand mobility all the way along the truss to some 24.4 metres above the Shuttle payload bay.

Crew Transfers & Loadmasters

In addition to the rotation of ISS resident crews and the delivery of new elements to install on the station, the main purpose of the Shuttle missions to ISS is the delivery of equipment, experiments and crew supplies for use inside the station. These 'Logistics Transfers' refer to the movement of items between the Shuttle and the station. The Russians gained extensive experience of this type of activity during the Salyut and Mir programmes, while the Americans limited experience of the intricate operations came during the nine Shuttle dockings to the Mir station under Phase 1.

Aboard the Spacehab module Ellen Ochoa, checklist in hand and Loadmaster for STS-96 spent many hours supervising the transfer of logisitics intended for the ISS.   (NASA)

The transfer of items between the vehicles is orchestrated by one of the crew, designated the Loadmaster, who serves as the lead for all transfer tasks, organises the unloading of the orbiter and Spacehab or MPLM, and supervises the preliminary stowage aboard the ISS or the location of items for return by Shuttle. The Loadmaster is the single point of contact between the crew and Mission Control and has responsibility for keeping all the transfer lists up to date. Depending upon the amount of items to be transferred, either one or two loadmasters are assigned, with one concentrating on the unloading and the other on stowing.

Loading the Shuttle during launch processing is an intricate operation that depends upon the type of logistics to be carried (perishable, delicate, dry, time critical, contaminants, etc.), their mass and volume (size and weight limitations), the sequence of unloading (which items have to be transferred on which day and in what order), the constraints of the mission design (EVA operations, crew exchange, element delivery) on crew time, and the launch constraints of the vehicle and landing contingencies for possible abort situations. The return of items also has to be carefully orchestrated to ensure that all items are safe for return, secured to prevent damage or injury once gravity takes over during landing, and whether experiment samples have to be located quickly. This takes months of planning and is prone to late changes as mission requirements change. The crew training programme also includes sequencing the method of transfer and stowage to be completed on the actual mission. The transfer time is sequenced in and around other activities and can include installation and power up of some transferred items.

Once docked to ISS, the transfer of items depends upon when access to ISS is planned, again around other activities and objectives. It has to be remembered that the hatches to and from the station are not always continuously open during the time the Shuttle is docked to the station. Items that can be transferred on any of the access days take secondary priority to ones that have to be transferred on specific days, and are fitted around the priority list accordingly.

Transfer facilities have been described under Spacehab and include Crew Transfer Bags, water bags, complete science racks and unique items (such as the treadmill). Stowage on ISS can be temporary or permanent and the whole crew is usually involved in changing items throughout the station. Certain locations cannot be used for stowage due to thermal, ventilation, mission safety, station control and crew access concerns. Items are securely restrained with Velcro straps and Russian bungees under flight rules governing ISS transfer and stowage constraints.

These constraints include not blocking the Zarya corridor, as this hinders the access to the Soyuz if the crew has to leave the station quickly; panels that require crew interface (fire holes and caution and warning panels can never be blocked by transfer items); and air circulation vents to prevent a recurrence of the pollution problems encountered by the STS-96 crew.

Transfer operations begin with the unbuckling of launch restraints or opening of lockers, and continues to the

Cosmonaut Yuri Gidzenko appears dwarfed by transient hardware aboard the Leonardo MPLM Note the amended checklists, colour coded labels (at left) and restraint straps holding the logisitics transfer bags in the stowage racks.   (NASA)

Once transferred into ISS finding the suitable stowage location initially takes a little while. Here STS-92 astronaut Michael Lopez-Alegria is almost lost in the far end of Zarya.   (NASA)

stowage and securing of items in the desired location. During the transfer, the Loadmaster refers to transfer lists to sequence and track the operation and keeps in contact with MCC to advise on progress and request advice if required. The information included in the transfer lists includes the sequence of transfer and item location; how the item is to be transferred (in a bag or stand-alone); and whether any items require special notes, handling, or reference to installation procedures.

Moving larger items such as this rack aboard Destiny Laboratory takes a couple of pairs of hands. It's not the mass of the object that's the problems it's stopping the movement as it moves and negotiating the hatches and corners without damaging the station, equipment or the astronauts.   (NASA)

Experience on the Russians stations found that personal items, fresh food, mail from home and other 'crew treats' were always some of the most quickly requested items to be transferred, and ISS is no different. Re-supply items can also include items stored in the orbiter middeck as well as those from Spacehab or MPLM. These normally include updated Station Operations Data Files and other documentation, new resident crew personal items and contingency water containers filled from the orbiter galley. When a new resident crew arrives by Shuttle, their individually moulded Soyuz seat liners need to be exchanged with those of the outgoing crew, and these are normally stowed in the MPLM. Items due for return include harmful contaminant filters from various locations and the former resident crew's seat liners.

At the end of the day's transfer, before or during a pre-sleep period, the Loadmaster transmits the number of items transferred during that day to the MCC duty Flight Activities Officer, appraising them of items stowed or not transferred and for what reason, or if the location of stowage had to be changed. Using a bar code information system helps track items, but it is only as good as the person who inputs the data and keeps the record current.

Undocking and Fly Around

Undocking from ISS has mainly occurred without incident and on time to allow the ISS crew to perform final preparations for the arrival of a new Shuttle, Soyuz or Progress vehicle. However, during STS-100, problems in qualifying the computer software caused controllers to keep the orbiter docked to the station for an extra day to allow additional time to review the computer systems on the US Command and Control Computer. This meant that Atlantis undocked less than one day before the first Soyuz Taxi mission (TM-32) docked at the station.

At the end of joint orbital activities, the Shuttle crew prepares the vehicle for undocking. Once the hatches have been sealed and verified, the latches are released and the initial separation is initiated by activating springs in the docking mechanism to gently push the Shuttle away from ISS. This prevents any thruster firing in close proximity to ISS elements. At 2 feet, the steering jets are activated once more. The undocking and separation manoeuvres and subsequent fly around profile are normally flown by the Shuttle pilot, essentially reversing the tasks performed by the commander during approach and docking and giving the pilot valuable experience for future dockings as a mission commander.

With the attitude control jets back in the 'Low Z' mode the pilot, working from the Aft Flight Deck, reverses the flight through the corridor cone to a distance of 450 feet, where the fly around begins. Initially, this starts directly behind and then directly underneath the station, progressing to above the ISS. Depending upon

A happy STS-102 crew, in a rather crowded middeck, photo document the separation from the station at the end of the joint activities, which on these missions saw the exchange of the first two resident crews.   (NASA)

available fuel reserves, two circuits of the ISS are flown, allowing the rest of the Shuttle crew to video and photo document and visually describe the exterior condition of the station and any new elements. As the Shuttle passes directly over the station for the second time, the pilot fires the control jets to complete the final separation burn. This places the departing Shuttle on a flight path that passes the ISS about 0.5 miles behind the station, then about 1.5 miles below it, before moving ahead of the complex and increasing the separation distance.

De-orbit and Landing

The final days in orbit are a return to the nominal Shuttle operations that have been performed by every returning crew since 1981. The crew configures the orbiter for landing, including stowing all unpacked equipment and setting up the middeck and aft flight deck MS seats. For the returning resident ISS crew they will take up a prone position on the Shuttle middeck with their head and torso parrallel to the middeck floor restrained on Recumbent Seat System Assemblies with thier feet in Vacant middeck lockers. These were first used during the Shuttle-Mir programme [2], to assist in re-acquaintance with 1g.

On several ISS Shuttle flights, small payload deployments were undertaken during this phase of the flight.

Mission	FD	Deployment
88	11	SAC-A
	12	Mighty Sat
96	10	STARSHINE
105	11	SIMPLESAT-1
108	12	STARSHINE-2

The flight deck crew completes an orbiter systems check, including control surface check and firing the RCS where necessary, depending upon their use during docked operations. During their last full day in orbit they receive weather reports on primary, secondary and contingency landing opportunities.

The crew usually spends half of their last full day in orbit off duty (Landing Minus One Day Activities), resting after a busy mission schedule and doing what all space travellers enjoy the most – looking out of the window at the Earth passing below. If the landing is waived off for the day, the crew spends time undoing their preparations for landing (termed De-Orbit Blackout), allowing them time to reconfigure the crew compartment for another 'night' in orbit and open the payload bay doors. They also use the extra time to communicate with their families by computer and generally enjoy the relaxation and fun of weightlessness.

Of the 12 missions reviewed in this article, the first four missions to ISS (STS-88, 96, 101 and 106) all success-

Towards the end of the mission the STS-108b crew deploy the Star Shine 2 small satellite. Note the MPLM at the rear of the payload bay, inaccessible by the crew when stowed on the shuttle.   (NASA)

fully landed on Runway 15 (N to S) at the Cape. Crosswinds cancelled the STS-92 landing attempts at the Cape on 22 October 2000 and on the next day, the high winds cancelled an attempt at the Cape, and rain and clouds aborted two attempts to bring the orbiter home at Edwards. The landing was finally achieved at Runway 22 at Edwards, after the previous 23 consecutive landings had been at the Cape.

The next mission, STS-97, was more successful at returning to the Cape, again landing on Runway 15. STS-98 had two landing opportunities on 18 and 19 February waived off due to weather concerns, as both were landing opportunities at the Cape. On February 20, the Shuttle took the first opportunity to land at Edwards AFB on Runway 22.

The first of two landing attempts for STS-102 was waived off due weather, but the Shuttle returned to SLF Runway 15 at KSC at the 2nd attempt. Bad weather also ruled out a landing in Florida for STS-100, so the orbiter landed at Runway 22 at Edwards. The STS-104 mission again had a weather waive off and stayed an extra day in orbit. The weather cleared the next day and the orbiter landed at the second attempt on Runway 15. It also took the second attempt on the first day to land STS-105 on Runway 22 at KSC, while STS-108 landed on the first attempt at the Cape on Runway 33 at KSC.

In summary, 9 missions returned to the KSC SLF, despite several problems with high winds and weather concerns, and three were diverted to Edwards.

Post Landing

Landing at the SLF at the Cape means the orbiter can quickly be returned to the OPF for the processing crew to service the it after the mission and conduct any repairs and modifications necessary before it begins the processing for its next mission. Landing at Edwards in California (STS-92, 98, 100) delays the processing, because the vehicle has to undergo a flight back to KSC atop the Boeing 747 Shuttle Carrier aircraft – at a cost of about $1 million each time. Once in the OPF, the support elements of the flown payload are removed, the Spacehab and MLPM are unloaded and serviced and the crew module and storage tanks are empted, purged and prepared for the next mission, where the cycle begins again. Meanwhile the crew return to JSC in Houston for a series of welcoming home ceremonies and de-briefing programme to evaluate the mission and help improve the training and preparation programme of future shuttle crews

Upcoming Shuttle Manifest

The success of the Shuttle missions during the period December 1998-December 2001 was reflected in the planning documents from 2002. Under the shadow of further budget cuts, mission planners hoped to fly four ISS-

At the end of the mission the crew thanks the families and workers at JSC in Welcoming Home ceremonies held at Ellington Field near JSC [copyright AIS photo] and inspire the support for the next mission already approaching the end of processing for launch back down at the Cape.
(Astro Info Service Collection)

related Shuttle flights each calendar year into 2007 (allowing for periods of orbiter maintenance down time, and the possible introduction of Columbia flying its first ISS missions). The majority of manifested Shuttle flights through to May 2004 were assembly flights aimed at completing the US Core [Ref 9]. The varied flight experiences of the 12 missions reviewed here give an indication of what to expect during the next dozen ISS missions, should all the varied and complex elements come together as planned.

Mission requirements determine the need for transportation both to and from ISS. This includes the number of astronauts aboard the ISS and their intended duration; the requirements for the life support systems; basic crew supplies (food, clothes, water etc.) per crew member times their duration, plus contingencies; the equipment for onboard propellant; the utilisation of the station and its facilities; hardware replacement, expansion and repair; and frequency of returned samples, logistics and waste. All these can be achieved with the combined use of the Shuttle and Soyuz, Progress and the Automated Transfer Vehicle. Currently the only method of crew rotation is by Soyuz or Shuttle. Re-supply can be achieved by Progress and ATV, and both can be used to incinerate waste upon entry at the end of the mission, but only the Shuttle is capable of returning large masses to Earth (max 25 tons, compared to the Soyuz mass of less than 50 kg).

It is clear that for the expected life of ISS, the NASA Space Shuttle fleet will demonstrate its value, finally confirming what George Mueller predicted at the BIS meeting over three decades ago: "Essential to the continuous operation of the space station will be the capability to re-supply expendables as well as to change and / or augment crews and laboratory equipment. Therefore, there is a requirement for an efficient Earth-to-orbit transportation system…(the) Space Shuttle."

Acknowledgements

In the preparation of this paper, the author expresses his gratitude for the continued assistance of: the staff of PAO Houston, Texas and at KSC. Jody Russell of the Still Photo Library, JSC, Margaret Persinger of the Still Photo Library at KSC; Shuttle Flight Directors Chuck Shaw and Glenda Laws; thanks to John Charles of NASA, JSC and Jeff Carr of USA, Richard Clifford of Boeing and Bernard Harris of Spacehab. In addition, the series of spaceflight articles by Roelf Schulling (Space Shuttle missions) and Neville Kidger (ISS operations) were frequently consulted. Thanks are also due to Rex Hall, Bert Vis, Mike Cassutt, and Mike Shayler.

References

1.	George E. Muller, "Manned Spaceflight: The Future", *Spaceflight,* **10**, pp.406-414, 1968.
2.	David J. Shayler, "American Flights to Mir (Space Shuttle)", *The History of Mir 1986-2000*, pp.71-85, The British Interplanetary Society, 2000.
3.	The Spaceflight Operations Contract: Five years of success

in space operations, briefing notes issued by United Space Alliance, 2002. with additional details provided during an AIS interview with Jeff Carr, Director of the USA Media Branch , Houston , Texas, May 10, 2002

4.   David J. Shayler, *Disasters and Accidents in Manned Spaceflight*, pp.145-149 , Springer-Praxis , 2000

5.   David J. Shayler, "The Proposed USSR Salyut and US Shuttle Docking Mission c.1981", *JBIS*, **44**, pp.553.562, 1991.

6.   David J. Shayler, *Disasters and Accidents in Manned Spaceflight*, pp.223-227, Springer-Praxis , 2000.

7.   NASA Space Station Freedom External Maintenance Task Team Final Report (2 Volumes) , William Fisher and Charles Price , 1990, NASA JSC.

8.   David Portree and Robert Trevino, Walking to Olympus: An EVA Chronology, Monographs in Aerospace History Series No. 7 , NASA, October 1997.

9.   Flight Assignment Working Group (FAWG) Planning Manifest, dated 10 June 2002.

Other References

Web Sites:

NASA Human Spaceflight Web Site (Space Shuttle and ISS) also the web sites links to JSC (Press Releases, Mission Status Reports) and KSC (Shuttle Countdown Online & History Archive) - http://spaceflight.nasa.gov

AIS Web Site - www.astroinfoservice.co.uk

Spaceflight Articles:

Roelof Schuiling articles (from Vol 41 March 1999 (STS-88) to Vol 44 March 2002 (STS-108) and Neville Kidger Chronology of ISS (from Vol. 42 Nov 2000 (July 2000) to Vol. 44 March 2002 (December 2001).

Reference Publications:

International Space Station Familiarization (TD9702) Mission Operations Directorate Space Flight Training Division, NASA JSC, December 1997.

Press Information Book, International Space Station, Mission Modules, Station Overview with updates Boeing, 1998-2000.

Reference Books:

David M. Harland, *The Mir Space Station: A precursor to space colonization*, **Wiley-Praxis 1997.**

David M. Harland, *The Space Shuttle, Roles, Missions and Accomplishments*, **Wiley-Praxis, 1998.**

Ernst Messerschmid & Reinhold Bertrand, *Space Station Systems and Utilization*, **Springer-Verlag, 1999.**

Dennis R. Jenkins, *Space Shuttle, The History of the National Space Transportation System. The first 100 missions*, **Midland Publishing, 2001.**

David M. Harland and John E. Catchpole, *Creating the International Space Station*, **Springer-Praxis, 2002.**

Peter Bond, *The Continuing Story of The International Space Station*, **Springer-Praxis ,2002.**

*   *   *

Research in Orbit

ANDY SALMON

Expedition 1 – Moving in

Science began on ISS even before Expedition 1. Logistics shuttle missions outfitting the stripped down Zvezda and Zarya modules included, as well as more mundane items like computer printer paper, experimental hardware for use by the Expedition 1 crew. The MACE-II technology test apparatus was on 2A.2B (STS-106) in September 2000 and the same flight delivered the EGN protein crystal growth canister to the Zarya module. The experiment was returned to Earth by shuttle mission 3A (STS-92) in October 2000, even before Expedition 1 had boarded ISS! It was a "risk mitigation" experiment to test out an enhancement to a technique for growing crystals.

Three Progress tanker/freighters (Progress M1-3, 1P, in August 2000, Progress M1-4, 2P, in November 2000 and Progress M44, 3P, in February 2001) brought a total of 114 kg of Russian scientific hardware for use by the Expedition 1 crew as well as supplies and propellant.

The three crew of Expedition 1 were on ISS to get the station up and running, particularly the Zvezda module. Science and technology tests would get in the way of the outfitting and so were kept to a minimum. Science operations weren't even intended to start this early but managers realised that small investigations could be carried out on a "non interference" basis. The crew workload was high but they still found time for some investigations (see Table 1). The figure given to a congressional committee in April 2001 was that with a crew of three only the time of half a person is available for research work.

Earth Observation studies began in November using the many windows of the Zvezda module. The MACE-II technological experiment had problems starting up on 9th January 2001 but Bill Shepherd got it working and left it onboard ISS for use by Expedition 2. Shepherd made first use of an essential tool for Expeditions: On Board Training. A CD-ROM for their laptop computers reminds them of the training they had with the payload developers a year or more before launch: payload overview, procedures and what it's trying to accomplish.

The EGN dewar was the perfect experiment during ISS construction so it flew again in January 2001. Tuck it away in the Zarya module and simply leave it for a month to thaw and grow crystals ready for collection later. Crew time required for it was an absolute minimum.

Sergei Krikalev and a long focal Earth observation camera in Zvezda.   (NASA)

TABLE 1: *Expedition 1 Science and Technology.*

EARTH OBSERVATION		
Crew Earth Observations	US	
Uragan (Hurricane)	Russian	
TECHNOLOGY		
Mid-deck Active Control Experiment II (MACE-II)	US	
MATERIALS SCIENCE		
Plazmenniy Kristall (Plasma Crystal, PKE-Nefedov)	Russian/German	Fundamental physics
PROTEIN CRYSTAL GROWTH		
Protein Crystal Growth-Enhanced Gaseous Nitrogen Dewar (PCG-EGN dewar)	US	
ISS ENVIRONMENT		
Identifikatsia (Identification)	Russian	Study of dynamic loads on ISS
Izgib (Bending)	Russian	Studying the microgravity environment of ISS
Tenzor	Russian	Measuring the dynamic characteristics of ISS
Iskazheniye (Distortion)	Russian	Measuring magnetic levels around ISS
Privyazka (Alignment)	Russian	Measuring ISS alignment characteristics
Spatial differences in CO2 concentration	US	
Treadmill Vibration Isolation System	US	Vibration reduction
MEDICAL		
Sprut-MBI (Octopus)	Russian	Study of fluids in the human body
Paradont	Russian	Human dental studies
Kardio-ODNT (using Chibis LBNP suit)	Russian	Study of cardiac activity and blood circulation in the human body
Interim Resistance Exercise Device	US	New exercise device
RADIATION		
Prognoz (Forecast)	Russian	Real-time radiation studies
Bradoz	Russian	Radiation biology studies
EDUCATION		
Space Exposed Experiment Designed for Students (SEEDS)	US	Plant Biology
Earth Knowledge Acquired by Middle School Students (EarthKAM)	US	Earth photography

Russian experiments concentrated on physiology and space station engineering studies with the notable exception of the Russian-German Plasma Crystal study. This was the first physical science experiment onboard ISS and in an important new area of fundamental science. It was delivered in February 2001 by the Progress M44 (3P) freighter and the Expedition 1 crew conducted the first experiment run in March 2001. Delays to Progress launches meant that not all the Russian experiments originally planned got underway during Expedition 1 [1].

Some of the work done was to characterise the ISS environment so that apparatus could be placed in the right locations or that data gathered could be corrected in the appropriate way. Despite the opportunity to do such tests on the ground there is no substitute for "all-up" testing once in orbit.

Some equipment was for operational medical studies rather than science – such as the monthly crew medicals or radiation safety monitoring.

A foretaste of what was to come was a 2nd February 2001 test link-up with ISS by the new Payload Operations Centre at the Marshall Spaceflight Centre and a preview of professional work to come was afforded by the educational science of plant seed studies by the Park seed company (providers of the tomato seeds for the LDEF shuttle free flyer) supported by the Jason program of Robert Ballard; and Earth observation photography scheduled via the internet using a camera for students set up in a Zvezda observation window. EarthKAM followed successful use on the Space Shuttle and was a project headed by former astronaut Sally Ride.

Expedition 2 – Destiny in Space

Expedition 1 successfully made ISS habitable and the delivery of the first set of large solar arrays meant there was now sufficient power to carry out research. The docking of Destiny, the research laboratory module, by 5A (STS-98) in February 2001, meant that the stage was set for science to begin in earnest by Expedition 2 (see Table 2). First

Destiny had to be fitted out. Its empty shell had only been provided with racks of ISS control systems by 5A and the next shuttle flight (5A.1 or STS-102) brought the first science rack, Human Research Facility rack 1, for human physiological studies in microgravity and the radiation environment of space.

A lot of the science required set up, activation and routine tending. Typical science tasks for the crew were: recharging sensor batteries, downloading sensor data to payload computers, checking experiment status panels and taking photographs of hardware set-ups. Repair tasks are also carried out – such as replacing a bent rod in the ARIS vibration cancellation system. Crew time for science was limited by the need for construction spacewalks, robot arm checkout, Quest airlock module activation and problems like control computer crashes or a misbehaving station robot arm. 24/7 operations by the Payload Operations Centre at the Marshall SpaceFlight centre were now underway for coordination of ISS science tasks.

The first commercial research began. The Commercial Generic Bioprocessing Apparatus was studying antibiotic production for pharmaceutical firm Bristol Myers Squibb. Sadly the apparatus, despite being space proven, shutdown and couldn't be restarted but it did fly successfully with Expedition 4. CGBA marked the first time commands were sent to ISS that originated at a remote location: from BioServe Space Technologies in Boulder, Colorado. The Advanced Astroculture (ADVASC) greenhouse was growing plants for a customer involved in science education. US and Russian/German commercial protein crystal growth units were also started up in Zvezda and Destiny, joining NASA protein crystal experiments in Destiny.

Yuri Usachev and a plant in the Zvezda module.   (NASA)

Physics experiments with the PKE plasma crystal apparatus continued and was now joined by the PCS experiment studying colloids (fine particles in a liquid). Scientists on the ground monitored PKE, PCS and ADVASC in an activity known as telescience. The Ku-band antenna on the Z1 truss, controlled by Destiny's computers, allowed a much higher volume of data (including video) to be returned to Earth through NASA's TDRS satellite network. Activation of some Expedition 2 payloads was delayed due to problems in starting operations with the Ku-band system.

The camera mounting rack for Destiny's optical window (WORF) wouldn't be ready until 2003 so Earth observation was by means of hand held still and video cameras.

Radiation monitoring involved a suite of sensors from Germany, Japan and NASA joining existing Russian sensors. Appropriately, 2nd April 2001 saw the largest solar flare for 25 years but fortunately it gave only minor increases to radiation levels inside ISS. The Canadian H-reflex biomedical experiment inaugurated contact with an ISS experiment from a centre outside USA/Russia – it was the Canadian Space Agency Telescience payload operations centre in Quebec. Susan Helms continued technology tests using the free-floating MACE apparatus in order to build steadier satellites in the future.

TABLE 2: *Expedition 2 Science and Technology.*

PLANT BIOLOGY		
ADVASC – Advanced Astroculture	US	
BIOLOGY		
Poligen	Russian	Fruit fly genetic studies
Oner	Kazakh	Seeds and fruit fly studies. During VC-1
ISS ENVIRONMENT		
Active Rack Isolation System	US	
Payload Equipment Restraint System	US	Tool and equipment personal storage
SAMS-II	US	Micro-g sensors
MAMS	US	Micro-g sensors
Identifikatsia (Identification)	Russian	Study of dynamic loads on ISS
Izgib (Bending)	Russian	Studying the microgravity environment of ISS
Tenzor	Russian	Measuring the dynamic characteristics of ISS
Iskazheniye (Distortion)	Russian	Measuring magnetic levels around ISS
Privyazka (Alignment)	Russian	Measuring ISS alignment characteristics
Infrazvuk-M	Russian	Acoustic mapping
Meteoroid	Russian	Micrometeorite and space debris monitoring
Vektor-T	Russian	Satellite navigation studies
Skorpion (SKR-1 equipment)	Russian	Experimental environmental conditions
RADIATION		
Bonner Ball Neutron Detector	Japan	
Dosmap – Dosimetric Mapping	German/US	
Phantom Torso	US	
Prognoz (Forecast)	Russian	
Bradoz	Russian	
EARTH OBSERVATION		
Crew Earth Observations	US	
Diatomeya	Russian	Ocean bio-productivity
Uragan (Hurricane)	Russian	
Zondirovaniye (Sounding)	Kazakh	During VC-1
BIOTECHNOLOGY		
Commercial Generic Bioprocessing Apparatus	US	Antibiotic production
PROTEIN CRYSTAL GROWTH		
Protein Crystal Growth – Biotechnology Ambient Generic	US	
Commercial Protein Crystallisation Facility-2	Russian/German	
Commercial Protein Crystal Growth – High Density	US	
Protein Crystal Growth – Single locker Thermal Enclosure System	US	
Enhanced Gaseous Nitrogen Dewar	US	
MEDICAL		
H-Reflex	Canadian	
Interactions	US	Psychological study
Sub-regional bone-loss	US	Pre and post flight only
Paradont	Russian	Human dental studies
Kardio-ODNT (using Chibis LBNP suit)	Russian	Study of cardiac activity and blood circulation in the human body
Farma (Pharma)	Russian	Effectiveness of medicines in micro-gravity
Dastarkan	Kazakh	Foodstuffs evaluation
MATERIALS SCIENCE		
Physics of Colloids in Space	US	
Plazmenniy Kristall	Russian/German	
TECHNOLOGY		
MACE-II	US	
Massoperenos (Mass Transfer)	Russian	Development of a new plant growth support system
EDUCATION		
Biosfera	US/Russian	Biology
EarthKAM	US	Earth photography

Mikhail Tyurin and Konstantin Kozeyev with the PKE apparatus. (NASA)

James Voss and the ADVASC experiment. (NASA)

Some of the exercise devices on ISS, like TVIS the treadmill, have fittings that reduce microgravity disturbances to onboard experiments but this was taken a step further with the ARIS fitting in a science rack. This equipment actively counteracts any disturbance it senses so as to provide a clean microgravity environment without "spikes" of acceleration.

There was just one Progress freighter/tanker visit during Expedition 2 (Progress M1-6, 4P, in May 2001) bringing 37 kg of scientific equipment. The Visiting Crew-1 in Soyuz TM-32 (2S) had another 9 kg of equipment with them. The flight to change the docked Soyuz lifeboat for a fresh one brought "spaceflight participant" Dennis Tito but the visiting crew also conducted experiments for Kazakhstan and took the opportunity to conduct more joint experiments with Expedition 2 whose results could be returned in the Soyuz capsule – items like video cassettes and PCMCIA cards.

Expedition 3 – Cells, Spacewalks and Visitors

Some Expedition 2 experiments that were complete were now on their way back to Earth, others continued operations and were joined by additional experiments (see Table 3).

TABLE 3: *Expedition 3 Science and Technology.*

BIOLOGY		
Aquarius	French	During VC-2
ISS ENVIRONMENT		
Active Rack Isolation System	US	
SAMS-II	US	Micro-g sensor
MAMS	US	Micro-g sensor
Identifikatsia (Identification)	Russian	Study of dynamic loads on ISS
Izgib (Bending)	Russian	Studying the microgravity environment of ISS
Tenzor	Russian	Measuring the dynamic characteristics of ISS
Izkazheniye (Distortion)	Russian	Measuring magnetic levels around ISS
Privyazka (Alignment)	Russian	Measuring ISS alignment characteristics
Meteoroid	Russian	Micrometeorite and space debris monitoring
Vektor-T	Russian	Satellite navigation studies
Skorpion (SKR-1 equipment)	Russian	Experimental environmental conditions
Kromka (Edge)	Russian	EVA deployed
Mirsupio	ESA	Experiment support tool
RADIATION		
Bonner Ball Neutron Detector	Japan	
Prognoz (Forecast)	Russian	
Bradoz	Russian	
EARTH OBSERVATION		
Crew Earth Observations	US	
Diatomeya	Russian	
Uragan (Hurricane)	Russian	
Immedias	French	Environmental studies
GEOPHYSICS		
LSO	French	
PROTEIN CRYSTAL GROWTH		
Dynamically Controlled Protein Crystal Growth	US	
Advanced Protein Crystallisation Facility	ESA	
Glycoproteid-K – using CPCF-2	Russian	
Mimetik-K using CPCF-2	Russian	
GCF (Granada Crystal box)	ESA	During VC-2
TISSUE GROWTH		
CBOSS	US	
MEDICAL		
H-Reflex	Canadian	
Interactions	US	Psychological studies
Sub-regional bone-loss	US	Pre and post flight only
Paradont	Russian	Human dental studies
Kardio-ODNT (using Chibis LBNP suit)	Russian	Study of cardiac activity and blood circulation in the human body
Farma (Pharma)	Russian	Effectiveness of medicines in micro-gravity
Sprut-MBI (Octopus)	Russian	Study of liquids in the human body
Diurez	Russian	Human metabolic studies
Profilaktika (Countermeasure)	Russian	Exercise studies
Cardioscience	Russian	Cardiovascular studies
Cogni	French	Human cognitive study
EAC	ESA	Psychological support
Effects of microgravity on pulmonary function (PuFF)	US	
Xenon-1 effects of microgravity	US	Blood pressure study; Pre and post flight only
Renal stone risk	US	Kidney stone countermeasures
MATERIALS SCIENCE		
Physics of Colloids in Space	US	
Plazmenniy Kristall	Russian/German	
TECHNOLOGY		
GTS	ESA/German	
HDTV	Japan	TV camera test

TABLE 3: *Expedition 3 Science and Technology (Contd).*

Dreamtime HDTV	US	TV camera test
SPICA-S	French	Electronic components
MISSE	US	EVA deployed
MPAC&SEED	Japan	EVA deployed
EDUCATION		
EarthKAM	US	Earth photography

The first significant science from a visiting crew was courtesy of the French space agency, CNES, in conjunction with the European Space Agency. Claudie Haigneré had a packed program during the brief joint operations with Expedition 3. Delivered by Progress freighter, by the Pirs module and by the Soyuz-TM capsule that brought her were a host of experiments. Claudie's crew continued PKE plasma crystal research as well as French biomedical tests that had long been a speciality of their flights to Mir. The visitors also brought the ESA Granada Crystal Box for protein crystal growth.

Spacewalk deployed experiments joined the fray for Expedition 3 after the arrival of both the NASA Quest and the Russian Pirs airlock modules. NASA had their MISSE long duration materials samples in containers attached outside the Quest airlock module. Japan had a similar experiment fixed outside Zvezda but with micrometeorite collection as well (MPAC&SEED). A Russian experiment, Kromka (Edge), studied the exhaust material deposited near thrusters outside the station. This caused much concern to EVA planners in case spacewalkers got the material on their suits and were exposed to it in the airlock – it is toxic. Spacewalkers now routinely carry dry wipes with them while on EVA – to remove any such contaminants on their suits.

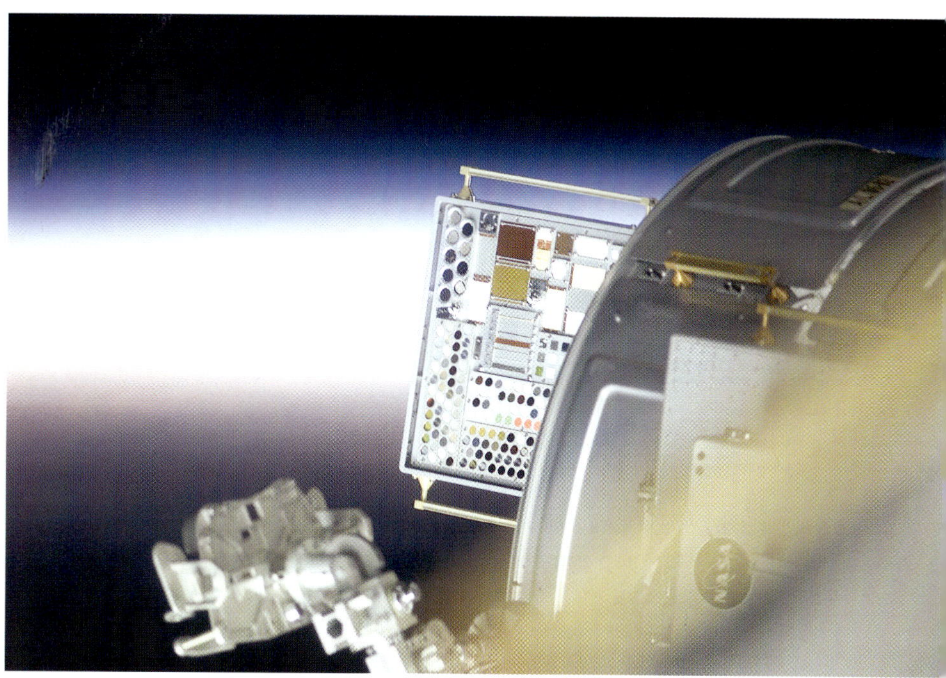

Backdrooped by a sunrise, the Materials International Space Station Experiment (MISSE) is visible on the International Space Station. (NASA)

The protein crystal growth experiments of Expedition 2 were superseded by the European APCF and the NASA dynamically controlled PCG unit. The latter was controlled from the ground, at the University of Alabama in Birmingham, with video monitoring of crystal growth and the capability to change the crystal growth rate. The German/Russian CPCF unit continued work.

Commercial technology testing got underway with the first steps in activation of the ESA/German Global Timing System. The CBOSS apparatus grew human breast and colon cancer cells along with kidney and nerve cells. This equipment took careful crew tending – from loading, to tending and monitoring, to preserving the results. Crew physiology studies included measures to reduce the risk of kidney stone formation in microgravity and detailed study of the lungs, including before and after spacewalks. The Physics of Colloids in Space experiment was conducting longer and longer experiment runs without the need for crew attention.

Scientific equipment for the Russian segment of ISS had a easier job getting to Expedition 3: two Progress tanker/freighters, the Soyuz TM of Visiting Crew-2 and even the Pirs docking module were loaded with a total of 314

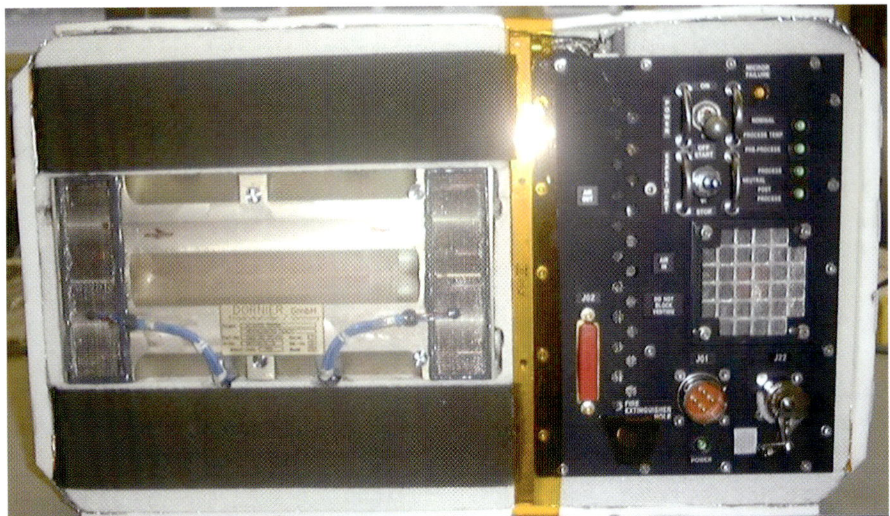

The Advanced Protein Crystallization Facility (APCF).   (NASA)

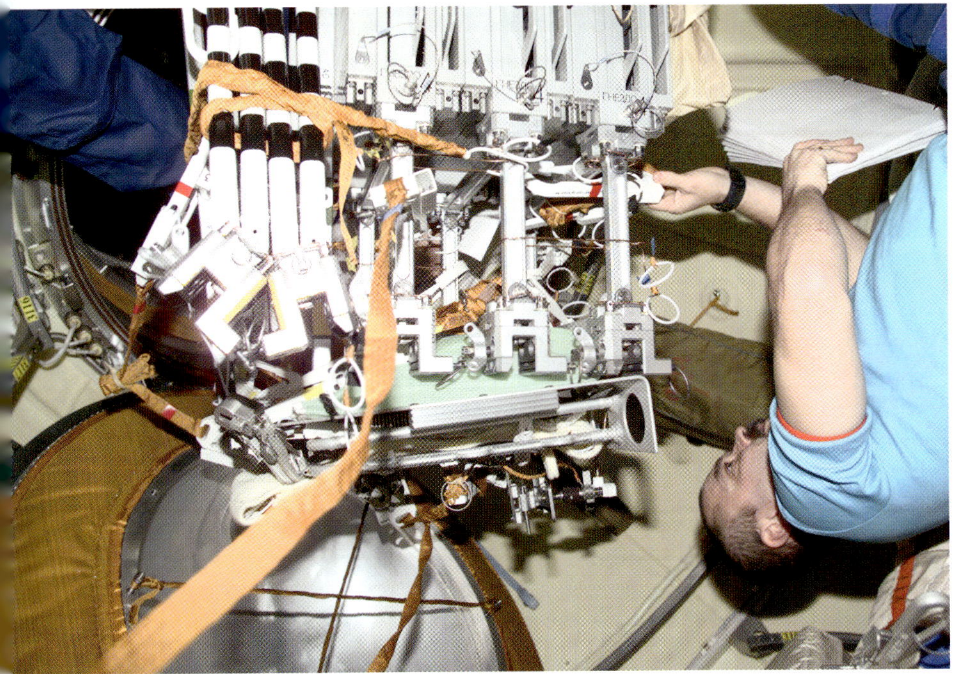

Mikhail Tyurin and the MPAC&SEED experiment before EVA deployment. (NASA)

kg of scientific items for delivery. Progress M-45 (5P) in August 2001 brought 150 kg from Japan (including MPAC&SEED for deployment outside ISS), 2 kg from France and 10 kg from Germany (the GTS electronic unit delayed due to lack of Progress cargo space). Pirs in September 2001 carried 47 kg of science apparatus from France and 4 kg from Russia. Soyuz TM-33 (3S) had another 15 kg of French scientific items. Progress M1-7 (6P) in November 2001 brought 33 kg of Russian scientific equipment.

Over the years 2000 to 2001 the International Space Station had grown from being a 70-ton apartment to a 150-ton house. 4 US astronauts and 5 Russian cosmonauts along with visiting crews had occupied it. The end result of early ISS operations was that it had living quarters, a laboratory, power and communications thanks to the hard work of the first three crews. What it needs most of all now is more crew time to tend to future experiments and make maximum use of its laboratory space because by early 2003 there could be 10 experiment research racks on orbit in Destiny alone. That's double the number present at the end of 2001.

Selected Experiments in Detail

These are some of the experiments or areas of science conducted onboard ISS.

MACE-II

The Mid-deck Active Control Experiment reflight is a follow-on to one that flew on the mid-deck of the shuttle in

Good Homes for Experiments

ISS reports are replete with names for equipment. Some are the names of experiment holders:

International Standard Payload Rack (ISPR) – the outer shell (the size of a kitchen fridge/freezer) for other apparatus. It will be used in all the ISS laboratories. The NASA Human Research Facility fills one ISPR

EXPRESS rack – one of the things that can go into an ISPR. It holds (for example) 8 "mid-deck locker" sized and 2 "drawer" sized compartments, all with independent temperature control, and can be equipped with the ARIS vibration prevention system

Drawer – based on one flown on Spacelab and Mir. It holds experiments

Mid-deck locker – as flown on the Shuttle mid-deck, Spacelab, Spacehab and Mir. It also holds experiments

By the end of 2001 there were 5 racks full of experiments on ISS: the Human Research Facility (a rack dedicated to biomedical research) and 4 EXPRESS racks with experiments from a range of sciences.

EXPRESS Racks readied for launch. (NASA)

1995. MACE-II was the first on-orbit demonstration of active control techniques, of self-monitoring and self-correcting systems, and also the first hands-on experiment on ISS. This means that it is both self-reliant and adaptive – smart enough to learn from experience, detect problems and fix itself without relying on a ground controller.

Satellites and spacecraft are designed and tested in gravity. It is very difficult to subtract the effects of gravity to predict how structures will behave in microgravity. If a satellite has large rigid structures it makes its behaviour easier to predict but this would be both massive and expensive. Space history has many examples of satellites and spacecraft that didn't behave in the manner expected due to large booms or other such structures that caused unwanted vibrations. Even the shuttle's robot arm spends half its time waiting for vibrations to die down when it's moved. MACE tests computer models of satellite behaviour in microgravity but also uses computer programs that are self-learning and which can react to new situations (be adaptive) without human intervention.

A 1.5 metre long structure has simulated instruments at each end with their own pointing systems. As one instrument is moved around on gimbals the software controls reaction wheels in the centre of the structure to keep the instrument at the other end of the structure pointing without disturbance. MACE-II was designed to allow intervention, almost in real-time, between the astronauts on ISS and experimenters on the ground. It can have its hardware reconfigured and new software loaded in the course of the experiments.

The experiment is from MIT and the Air Force Research Laboratory. The technology will find use on both civil and military satellites - any lightweight satellite that needs to keep pointed at a target of interest.

14 experiment runs were made during Expedition 1 and a further 62 on Expedition 2, each 2-90 minutes long. It was originally only intended for the first expedition and was set up in the Unity module, with a rigging of tethers, but there were problems in its first use in January 2001. The successes demonstrated when it got working (after a PC card was changed on 25th January) meant it was retained onboard and Expedition 2 trained in its use as well.

The large volume of data gathered by a laptop computer was transmitted to Earth using the Ku-band antenna on Expedition 2 – much better than having to wait for physical disk drives to be returned from Expedition 1.

Plasma Crystal Experiment
(Plasma Kristall Experiment, PKE-Nefedov [2])

The PKE apparatus studies a new state of matter: plasma crystals. The phenomenon was only discovered in 1994. It results from adding tiny plastic polymer spheres to plasma in a vacuum chamber to make "complex plasma". The plastic polymer spheres become negatively charged and behave as though they were atoms in a crystal lattice but they are large enough to be filmed and move much more slowly than atoms. When the plasma crystals are produced

Sortie Missions

Sortie payloads are sponsored by the ISS program to test equipment or experiments for later use on ISS or for those which don't need long duration flight. Several were flown during these early stages of ISS construction, piggybacked on the construction/logistics flights themselves. They never left the confines of the orbiter and spent about a week in space but they serve a purpose once reserved for Spacehab or Spacelab flights.

Mission	Experiment
2A.2B (STS-106)	Commercial Generic Bioprocessing Apparatus
7A.1 (STS-105)	Microgravity Smouldering Combustion
UF1 (STS-108)	Avian Development Facility
	Collisions into Dust –2
	Commercial Biomedical Testing Module
	Microgravity Smouldering Combustion
	Prototype Synchrotron Radiation Detector

in gravity, sedimentation means that only flat plasma crystals can form but in microgravity 3-D crystals are produced. Plasmas full of charged dust particles are an important part of objects as diverse as Saturn's rings, comet dust tails and proto-planetary disks around stars.

The Max Planck Institute for Extraterrestrial Physics (MPE) from Germany flew plasma crystal experiments twice on Texus sounding rockets from Sweden. The Russian Institute for High Energy Density (IHED) of Moscow flew their apparatus on Mir in 1998. PKE was the joint experiment taken to ISS in February 2001 for the study of dusty plasmas – the basic principles of fluid physics. It's possible to produce solid, liquid and gaseous states in dusty plasmas and thus to study transition phases, like melting, in detail. In microgravity the melting process can be studied in detail – something that is complicated by pressure forces due to gravity in a 2-D plasma crystal on Earth. PKE was operated by the Russians on the Expedition 1 crew and by the Visiting Crews 1 and 2 inside the transfer compartment of the Zvezda module. Expedition 1 carried out basic tests of the apparatus, varying gas pressure, radio frequency power and particle sizes, while the visiting crews studied particular effects such as waves in the plasma.

After the death of Anatoly Nefedov in February 2001, it was decided to name the experiment after him. He was head of IHED, leader on the Russian side of the experiment and a former science advisor for President Yeltsin. PKE uses a low temperature vacuum chamber with some low-pressure argon gas excited by a radio-frequency discharge. The vacuum is achieved by a direct connection to space. The unit was not achieving a perfect vacuum so a commercial off-the-shelf molecular pump was ruggedised and space qualified by the German company Kayser-Threde, taken up to ISS onboard a Progress supply craft and fitted before Visiting Crew-2 arrived.

The chamber is seeded with polymer spheres just 3 and 7 microns in size (less than the thickness of a human hair) and the plasma crystals are filmed by cameras when illuminated by lasers. Telescience controls mean that the conditions can be varied from the ground and quick results monitored as the experiment proceeds though this was not possible for Expedition 1 due to the lack of a Russian data relay satellite. The crew had to carry out experiment runs themselves under the control of the attached computer. Data is recorded onboard ISS, in the form of videos and PCMCIA card digital data, which are then returned to Earth by the Shuttle (STS-102 with Expedition 1) or in the Soyuz of a visiting crew. The Expedition 1 crew e-mailed some data to Earth while they were carrying out the experiments. Quick look video data was available to scientists on the ground during later experiments.

Advanced Astroculture (ADVASC)

The main plant growth experiments of recent years used the Svet greenhouse on the Mir space station. Astroculture was a NASA commercial unit developed for space shuttle use that also saw time on Mir. Advanced Astroculture is a further development for use on ISS. The University of Wisconsin-Madison's WCSAR commercial space centre and its customer Space Explorers Inc. grew Thale Cress (Arabidopsis thaliana) on ISS with a unit sitting in an Express rack in the Destiny module. This is a very well characterised and fast growing, small plant. The seed to seed cycle takes just 6-7 weeks. The objectives in this case were:

a) To see whether seed to seed life cycle could be followed in space: germination, growth and seed development. Seeds harvested from the experiment back on the ground would be used to grow new plants on Expedition 4. A. thaliana were grown in this manner onboard Mir but with no atmospheric control system so ethylene build-up caused growth problems

b) To use DNA analysis to study gene expressions in the plants – genetic plant traits. A freezer on Expedition 4 would allow tissue sampling and storage in the course of the experiment

c) To chemically analyse the seeds produced

ADVASC controls temperature, humidity, atmosphere, light, water and nutrient delivery. Seeds are planted in a peat moss like substrate and reservoirs are filled with water and nutrient while on the ground. The crew activated the unit when in space and the plants then germinated. The crew sampled nutrients, condensate and gases several times over the growth period. The crew regularly checked status displays on the unit. Telemetry and video was also sent to the WCSAR by the TRek telescience system so plant growth could be monitored on the ground. After activation on 10th May, the seeds had sprouted by 28th May and were flowering by 11th June. Seeds were produced within 2 weeks. The crew replenished the nutrient solution during the experiment. Preparation for return to Earth involved commands from the ground and actions by the crew. Nutrients, fluids and gases were removed from the growth chamber, the temperature increased and the humidity lowered to dry the seeds. The experiment returned on STS-104 rather than STS-105 because the plants had finished their growth cycle and both shuttle flights had been delayed due to Space Station robot arm problems. On this flight 90% of the seeds germinated and 70% produced seedpods.

Jim Voss performs most of the activities for the Advanced Astroculture experiment (ADVASC).
(NASA)

Huntsville Ops

The Payload Operations Integration Centre (POIC) in Huntsville, Alabama is now added to the familiar Mission Control (MCC-H) in Houston and TsUP (the Flight Control Centre in Korolev near Moscow). The site was used for controlling science on Spacelab but now has to manage the science operations for 3-4 times more experiments at any time. It operates 24 hours a day, 7 days a week and is staffed by 3 shifts of 6-19 flight controllers.

Its job is to plan the science for the Destiny laboratory, watch out for safety, handle science communications with the crew, manage payload resources, troubleshoot; and look after both commands sent to the payloads and data received from them. The POIC is linked to other centres and universities around the world who run experiments on ISS.

The Russian TsUP is responsible for the planning and operating of experiments in Russian modules but POIC synchronises payload activities, assures safety and optimises the use of resources.

Lybrease Woodard, a POIC POD (a Payload Operations Director, equivalent to a mission control flight director), described it as "like leading a new orchestra. Everybody learned how to play their parts and then came together like a symphony." Director of Marshall's Flight Projects Directorate is former shuttle astronaut Jan Davis.

Shakin' All Over

In orbit you don't experience "pure" weightlessness as you fall towards Earth as fast as your orbit takes you around it. What you actually get is something now termed microgravity – in other words forces that are 1/1000th that of one gravity. The two biggest effects felt within ISS are: drag forces from the tenuous outer atmosphere of Earth and gravity gradient effects (due to the slightly different orbits that parts of ISS not at the centre of mass follow around the Earth). Most of Destiny should experience micro-gravity levels of 1 micro-g, parts of Destiny and most of Zvezda experiences levels of 2 micro-g. Add to this disturbances from fans and pumps, the crew moving around or exercising, other craft docking or undocking, water dumps, thrusters firing, orbital re-boost from larger engines or airlock operations.

NASA and Russia take great care to measure micro-g levels onboard ISS. NASA data is available to researchers so disturbances can at least be factored into experimental results. The Glenn Research Centre specialised in flying sensors on the shuttle and Mir. It is doing the same for ISS: MAMS and SAMS-II.

SAMS-II looks at the disturbances caused by ISS, its crew and equipment such as exercising or fans. These are vibrations; from 0.01 to 300 Hz. SAMS-II uses sensors spread throughout Destiny. SAMS data, not surprisingly, showed transient disturbances due to the Temporary Sleep Station in the Destiny laboratory being used by a crewmember.

MAMS looks at bigger effects: the drag experienced by the whole of ISS plus the effects of ISS rotation and water dumps. These happen fairly slowly, on a timescale greater than 100 seconds, and are called quasi-steady accelerations. MAMS uses a spare sensor from the Shuttle OARE experiment.

Zvezda has accelerometers that are used in Russian studies (like Identifikatsia). The Skorpion experiment (SKR-1) was used inside ISS to measure the full spectrum of microgravity levels, radiation, electromagnetic interference and climatic conditions.

Further study of the dynamics of ISS is essential because it may have an adverse effect on thermodynamics or diffusion research planned for the station.

Radiation Detectors [3]

ISS orbits the Earth below the van Allen radiation belts but its orbit takes it through the South Atlantic Anomaly, where the van Allen belt of trapped radiation dips down towards the Earth. The ISS orbit is 51.6 degrees and this takes it near parts of the ionosphere where the magnetic field lines dip down towards the geomagnetic poles – it directly passes through radiation coming from the Sun during geomagnetic storms that push the auroral zones towards the equator. Radiation monitoring is both an operational requirement and an area of scientific interest: to research the effects of radiation on humans and other organisms in space.

Radiation has three sources:

Galactic Cosmic Rays – these are high energy particles from supernova explosions far off in the universe and the heaviest of these ions do most damage to the human body. GCR levels are highest at solar minimum (when there is least activity on the Sun) because the solar wind sweeps the inner solar system clear of many GCRs.

Van Allen Belt – the South Atlantic Anomaly is the manifestation of the van Allen belt experienced by spacecraft in low Earth orbit – ISS passes through the SAA about 5 times per day, with each passage taking about 25 minutes.

Solar Particle Events – SPEs arise from solar flares on the Sun. They normally last for several days to a week and do more damage to the human body than the brief passage through the South Atlantic Anomaly (where the body has time to make repairs after passage).

Radiation dose levels inside ISS are believed to be of less concern than those for spacewalkers caught outside during a Solar flare.

The ISS kayuta (sleeping compartment) was protected from radiation by high density polyethylene slabs and the temporary sleep station for the third crew-member (in Destiny) is shielded with both the slabs and bags of water for the same reason – lots of hydrogen atoms moderate the radiation. The Personal Radiation Protection System was developed and delivered by JSC in less than a month. The POLY-bricks (sheets of polythene) were delivered on the shuttle along with the Expedition 2 crew. The Zvezda slabs were later removed in March 2002 because Russian specialists, who feared secondary radiation from them, had not approved them.

Virtual Crew

The ISS is currently limited to a crew of three people. Their main task is construction work to assemble the ISS and maintenance work to keep it operating smoothly. Crew time for doing work on scientific experiments is limited to about 20 hours per week (for NASA experiments – add in a little more time for Russian experiments).

Telescience helps the situation. The crew start an experiment going but operations and monitoring is from the ground. NASA has set up a series of Telescience Support Centres for this purpose and control of NASA payloads is delegated to them:

Johnson Space Centre for human life sciences and physiology

Ames Research Centre for plant and animal studies

Glenn Research Centre for combustion and fluid science

Marshall Space Flight Centre for materials science and biotechnology.

NASA PC software called TreK allows telescience from any investigator's home base (termed a "remote site"). The availability of high data rate Ku-band communications now allows video as well as radio links. In conjunction with a Telescience Support Centre they can control their experiment and gather data.

So there are telescience sites around the USA and Canada, for example: in Canada for the Hoffman Reflex experiment, in Cleveland (Ohio) and Harvard University for the Processing of Colloids in Space, in Birmingham (Alabama) for the Dynamic PCG experiment and in Boulder (Colorado) for the Commercial Generic Bioprocessing Apparatus.

The first three expeditions accumulated 500 person-hours of crew time on experiments but there was almost 50,000 hours of telescience operations.

The main science operations centre is at the Marshall SpaceFlight Centre – where the Payload Racks Officer terms themselves "the fourth crew member". They monitor and configure the resources and environment for experiments. You could consider the whole of the POIC flight control team as one or more virtual crewmembers on ISS, with 100% of their time available for science operations.

Capcom is the voice of mission control in Houston – for command and control of ISS. But Paycom is the voice of POIC in Huntsville – handling voice communications between the crew of ISS and the POIC.

The Payload Rack Officer (POC) position at the Payload Operations Center.   (NASA)

A plethora of radiation sensors are on ISS, some deployed from the start of operations and some during particular expeditions, especially during Expedition 2. They measure different types of radiation. Active sensors give real-time read-outs but passive sensors have to be brought back to Earth for analysis. Real time measurements give instantaneous radiation levels and dosimeters give total doses over a long period. Some sensors, especially the passive ones, are returned to Earth on shuttle or Soyuz flights. For example STS-105 brought back the Phantom Torso and DOSMAP while STS-108 carried Bonner Ball home.

Manual interventions by the crew included daily status checks, a hard drive change out on BBND due to an error indicator, saving sensor data to computer using portable hard drives (to allow more to be gathered), recharging the batteries in portable radiation sensors and changing sensor operational mode after a solar flare (to gather data at a faster rate). Some computer data was sent to Earth on a regular basis and the rest was stored on hard disk for physical return to Earth.

The radiation sensors are:

Bonner Ball Neutron detector – a Japanese instrument providing real-time measurement of neutron radiation. Neutrons comprise about 20% of radiation near the Earth and can cause significant damage to blood forming bone marrow.

DOSMAP – a set of German, US and Hungarian real-time sensors and dosimeters (the latter are analysed upon return to Earth) in a German led experiment laid out in several ISS locations. The Hungarian sensors are Pille thermoluminescent dosimeters – earlier versions flew on Salyut and Mir.

Planning for the Future

An overall plan for the whole of an expedition is worked out before launch. This uses all the information from the experimenters as input and the end product is the Integrated On Orbit Summary (IOOS). Every week during the expedition, two weeks in advance, a Weekly Look ahead Plan (WLP) is produced – an updated and cut down version of the IOOS. There is just one WLP for each week – with both US and Russian input. Every day, four days in advance, a more detailed daily version of the WLP is prepared. This final timeline is uplinked to the crew on ISS and becomes the Onboard Short Term Plan (OSTP). At the same time TsUP in Russia uplinks the daily Form 24 (as used on Mir) listing the Russian crew activities. Effort is invested in getting the OSTP and Form 24 to match but they don't always.

A task list is also available for use by expeditions. This is the "Job jar" of Bill Shepherd on Expedition 1. It comprises non-time critical items that the crew can do if they complete other activities early or for off duty time. It is often viewed as a "get ahead list", for example activities planned for the next week may be put on the task list for the current week.

James Voss and the DOSMAP experiment power unit. (NASA)

Bradoz – a set of Russian Thermoluminescent and solid track (tissue equivalent plastic) dosimeters along with packets of seeds.

R-16 – a Russian dosimeter built into ISS for measuring ionising radiation (used in the Russian Prognoz experiment).

Phantom Torso – a head and torso with both real-time sensors and total dosimeters inside. The bones are real but the organs inside are made of material that simulates the real thing. This NASA experiment is for comparing actual doses received inside the body – which cause the real damage to humans – against the radiation levels that are actually measured – those on the skin. The head had flown on shuttle missions in the mid-deck area and the whole construct flew to Mir on STS-91. "Fred" the Phantom torso has 416 passive lithium crystal detectors and 5 active detectors and is the size of a normal male torso and head. Scientists are especially interested in radiation levels inside organs that create blood. Two passive detectors are carried in the shirt pocket of the torso for the "skin" radiation measurements to compare with all the internal sensors.

Early results from the experiment show that skin measurements are an accurate guide to internal radiation doses absorbed by internal organs for Galactic Cosmic Rays but that skin measurements are an overestimate of the internal doses caused by protons.

The last two detectors are part of the Crew HealthCare Systems (CHeCS) Environmental Health System on ISS:

Tissue Equivalent Proportional Counter – a NASA instrument that was set up by Expedition 1 after concerns about high solar activity in November 2000 which gave rise to the crew being told to shelter in the rear of Zvezda. It was also set up near the Phantom Torso experiment. TEPC flew on the Space Shuttle, on Mir, on the doomed Mars-96 probe and on the Mars Odyssey probe. Expedition 1 crew were subjected to about 30-40 millirem per day (about 3-4 chest X-rays).

The TEPC alarm was triggered on 4th November 2001 after a large solar flare. The crew received 80 millirads of radiation that day rather than the usual 20 millirads for Expedition 3.

James Voss with the BBND experiment and the Human Research Facility.　　　　　　　　　　　**(NASA)**

Phantom Torso.　　　　　　　　　　**(NASA)**

Calling Occupants of ISS

Communications with researchers on the ground can sometimes make or break an experiment. The crew get refresher training CD-ROMs for their laptop computers aboard but they need someone to check with if things don't go as planned.

Frank Culbertson, Expedition 3 commander, said that space/ground communications is one of the top three priorities for any ISS crew (along with time management and stowage management).

Early communications on ISS relied on Russian VHF communications links and the NASA S-band Early Communications System (Ecomm or ECS) installed in Unity and using NASA Tracking and Data Relay Satellites. It was decommissioned on a 6A spacewalk. Russian ground site communications passes lasted 10-15 minutes (20 minutes at best) and there were large parts of the day with no passes at all. NASA supplemented with US ground sites (as they had during Mir operations). The OCA videoconference capability was set up in November 2000 (on 4A). It provided low resolution, slow scan, Internet style videoconference for use by MCC-H and TsUP in Moscow using a laptop.

Eagerly awaited was use of the Ku-band antenna on the Z1 truss, which was delivered on 3A in October 2000 but not activated until the Expedition 2 crew were aboard. It is similar to one carried by the space shuttle orbiter and is a link to TDRS satellites in geostationary orbit. Ku-band is used for high data rate experiment data and video. It has 250 times the capacity (bandwidth) of S-band. S-band is used for voice communications, commanding and low data rate transmissions. Even with TDRS satellite communications there is a Zone Of Exclusion where there is no coverage or ISS attitude may prohibit line of sight contact with a satellite. TDRS coverage is normally for 60 minutes out of every 90 minute long orbit but with a 2-minute disruption every orbit when communications are handed over between satellites.

The Russian segment on ISS has its own steer able antenna (like the one on Mir). Eventually there will be Russian communication satellites available for ISS use – Luch is their equivalent of TDRS. One had been planned for 2000 then 2001 but didn't materialise.

IntraVehicular Charged Particle Directional Spectrometer – another NASA instrument that was also set up near the Phantom Torso experiment. It could be moved around modules to measure radiation coming from particular directions.

Crew Earth Observations

The Expedition 1 crew began taking Earth Observation images using the many windows in the Zvezda module. A porthole in Zvezda has enhanced optical quality but transmits dangerous ultra-violet radiation. Observation of the Earth's surface should only be made while ISS is in LVLH attitude (to prevent the chance of directly looking at the Sun) and sunglasses should always be used.

NASA gathered data for its human spaceflight Earth photography database [4]. This allows study of changes over many years since it goes back to 1961. The main topics are:

Human expansion and its environmental impacts

Seasonal changes

Sudden events like fires, floods, storms and volcanic eruptions

Target opportunities are sent up on a weekly basis (except when the station's orbit is changed for a re-boost) – to be fitted into the schedule as time permits and taking about 10 minutes per day. Crew chosen targets are also included. The Russian crew have similar programs – specifically: Uragan, studying areas of the world where natural disasters have/may happen and Diatomeya, studying bio productive areas in the world's oceans. The main difficulty encountered by Expedition 1 was to know exactly where they were above the Earth! Several planned targets were missed because they had already been over flown at the time they were due to make observations. New computer software to aid the process was delivered to ISS laptop computers.

The arrival of Destiny meant that the crew could now use the optically perfect 50 cm diameter, fused silica observation window in the laboratory. This is probably the highest quality spacecraft window ever. The window normally faces "down" towards Earth (nadir pointing) and had been fitted with the best glass through which observations could be made, rather than the shuttle windows, which were derived from X-15 pilot windows. There are 4 panes of glass but with anti-reflection coatings and wide spectral transmittance. The window even had replaceable panes on the inside and outside – so damage could be overcome whether debris/micrometeorite on the

Susan Helms lookng out of the Destiny lab window. (NASA)

outside or scratches on the inside. The innermost ("kick") pane has an ultra-violet and infrared coating to protect the crew. UV can give an observer a severe suntan if you spend too long observing and there is a danger to the eyes.

The "kick pane" was found to have numerous scratches and smudges upon examination in February 2002 – using a torch and a night sky background.

The main constraint on window use is that the outside shutter (fitted during a February 2001 EVA) must be closed by hand cranking when the ISS z-axis is within 85 degrees of the flight direction (to avoid contamination) or if the SSRMS robot arm is parked nearby. This means more than half the time when ISS is in XPOP attitude. It should also be closed whenever another craft manoeuvres near ISS. The window outside cover was left open during a 2001 Soyuz docking – leading to some concern about contamination build-up from engine combustion products.

Micrometeoroid/orbital debris (MMOD) damage to an Earth facing window in Zvezda by the end of 2001 showed that protective covers are useful – this 23 cm diameter window only had a shade rather than a cover.

A full frame mounting to fit inside the window was due to be delivered in 2003, the Window Observation Research Facility (WORF), so in the meantime the crew used hand-held still and video cameras aimed through the window with a variety of lenses (from 50 mm up to 800 mm). Binoculars of x20 magnification, with stabilisation, could be used to help identify targets. The availability of electronic still cameras meant that results could be transmitted to

A Matter of Attitude

ISS travels through space in one of two attitudes, XPOP or LVLH, and this can affect some of the experimental work being done.

XPOP means x-axis perpendicular to orbit plane (AKA "sun fixed").

LVLH TEA means local vertical, local horizontal, torque equilibrium attitude (AKA "Earth fixed"). This is how an aircraft normally flies (z-axis down, x-axis pointing in direction of travel and right wing level). In this attitude ISS rotates once as it circles the Earth – in comparison with a fixed reference frame attitude like XPOP. Torque equilibrium attitude is –7 degrees yaw, -10 degrees pitch, which provides a stable gravity gradient attitude with minimal stress on ISS components.

Until there were more solar arrays on ISS there were times of the year when the Sun angle (due to the seasons) didn't provide enough solar power so the station attitude alternated between XPOP and the more usual LVLH for two weeks at a time. Some experiments needed LVLH in order to operate.

Earth soon after they were taken and that the crew could check their image gathering technique as well. Expeditions 1-3 gathered more than 13,000 images with 35mm, 70mm and digital cameras. The images had a surprising resolution of up to 6 metres [5]. The crew learnt how to compensate for the motion of ISS by tracking the ground targets below when using high magnification lenses.

Protein Crystal Growth [6]

Proteins are a key component of the cells that comprise life. Learning about the structure of the proteins (their "shape") is vital for both medical research and the pharmaceutical industry. The shape of a protein molecule allows it to act as either a "key" or a "lock" when matched with another "lock" or "key". The best way of doing structure analysis is to grow crystals from solutions of the proteins. The crystals are then analysed – using X-rays to produce diffraction patterns. Better crystals can be grown in microgravity. To begin with, larger crystals could be grown than in gravity but the use of X-ray synchrotrons for high-energy X-ray sources makes this less important today. Well-structured crystals are still needed to get fine detail structures and this is where microgravity crystals offer the most potential.

As shown by payload specialists on Spacelab flights, real-time monitoring of crystal growth can be beneficial and one of the ISS crystal growth experiments makes full use of this – but from the ground rather than onboard.

ISS research studies the relationship between experimental variables and crystal characteristics. This is fundamental research into macromolecular growth that is giving new techniques for use on Earth as well. Long-term crystal growth on ISS gives the chance to show whether crystals grown in microgravity have the potential to solve challenging problems in biology.

A variety of different techniques to grow crystals are being used on ISS, giving researchers the chance to see what best meets their needs:

Vapour diffusion (hanging drop) – water slowly migrates away from a droplet of mixed protein and precipitant at the tip of a syringe after activation of the unit. This is in a chamber surrounded by material soaked in an agent that attracts the water. As the hanging drop gets more concentrated a crystal grows. This method has temporarily been abandoned by ESA due to convection effects caused by surface tension in the hanging drop.

Vapour diffusion (Sitting drop) – a drop of mixed protein solution and precipitant is in a sample well. A surrounding moat of absorbent fluid draws water away from the drop and crystals then form.

Vapour diffusion (Porous plug) – a porous material separates the protein solution and a precipitant solution. Changes begin the moment the unit is loaded with solutions but it takes several days for noticeable effects – ample time for a shuttle to reach ISS.

Liquid-liquid diffusion (free interface diffusion) – a protein solution and precipitating agent are separated by a buffer solution and the precipitant slowly diffuses into the protein solution through the buffer once the unit is activated. This makes the protein less soluble and a crystal forms. The first ever space crystal growth experiment (by University of Freiburg on Spacelab 1) used this method.

Dialysis – a thin membrane, through which the agent slowly diffuses into the protein solution after the unit is activated, separates a protein solution and a precipitating agent. This causes crystals to begin forming.

Batch – a protein solution that is not quite saturated is mixed with a reagent after launch that alters the saturation level so it becomes saturated and a crystal starts to grow.

Conditions have to be just right or an amorphous mass rather than a crystal results.

The crew transferred the apparatus from a Progress freighter or from the shuttle's Spacehab Mid-deck or MPLM logistics module. The PCG units are the first things to be unloaded from a Progress supply craft after it docks. Some of the protein solutions degrade fast after loading into their growth chambers so they are put into the shuttle mid-deck just prior to launch up to ISS.

Activation is by the crew turning a screw or throwing a switch to bring solutions into contact with each other. Some units don't need activation – they are already starting to develop the conditions to form crystals after loading back on Earth – with a very slow melting or diffusion of vapours.

The crew made daily status checks of the PCG apparatus: checking the temperature.

The final transfer back to the Shuttle was critical. Some powered experiments (for temperature control) had only a 30-minute window available for the transfer but it was accomplished in about 5 minutes.

PCG Devices

The Commercial PCG high-density apparatus grew more than 1,000 crystals at a time (a big increase over previous space crystal growth units). It used a Commercial Refrigerator/Incubator Module (CRIM) for temperature control and the "sitting drop" crystal growth method.

The European Advanced Protein Crystallisation Facility carried 48 crystal growth experiments at a time. Growth reactors had protein volumes individually tailored to experiment needs. 10 of these could be monitored by video stills in 3-D. An interferometer studies 5 of the cells to look at things like concentration gradients. Pictures and interferograms are stored on tape for analysis back on Earth. Experiments in it can use either batch, hanging drop, liquid-liquid diffusion or dialysis (the latter two were used on Expedition 3).Experiments in it can use either liquid-liquid diffusion or dialysis (the latter was used on Expedition 3). Flight of APCF on ISS is courtesy of the Italian Space Agency – NASA gave them payload space in return for providing the ISS MPLM shuttle cargo carrier modules. ASI gave the space to ESA. APCF was used from August to December 2001 as the first ESA experiment operating aboard ISS.

The Dynamically Controlled PCG facility carried out 38 experiments at a time by the hanging drop method – which could all be monitored on live video and their growth controlled from the ground in real-time to refine the hardware and methods of protein crystal growth for future experiments. Dry nitrogen gas controlled the evaporation rate of the solution around the crystals. When problems developed with the video system in September the ground control centre found a work-around and transmitted commands that got half of it under observation again.

The PCG-STES unit (containing Protein Crystallisation Apparatus for Microgravity, PCAM, cylinders) grew 378 crystals at a time. Single locker Thermal Enclosure System (STES) provides a temperature-controlled environment and PCAM uses the "sitting drop" method of crystallisation. The unit on Expedition 2 only grew crystals in 24% of its chambers. The same thing happened in the ground control experiment so it was put down to the late activation of the experiment (16 days after being loaded in the growth chambers) and a similar experiment on Expedition 4 was to be activated sooner.

The PCG-BAG unit contained a single Diffusion controlled Crystallisation Apparatus for Microgravity (DCAM) tray in a bag. This uses the vapour diffusion (porous plug) method. It did not require power or temperature control just stowage space.

CPCF-2 was the first European experiment to be operated onboard ISS. It was a commercial protein crystal growth unit developed by Astrium for the German space agency, DLR, and their main customers: European pharmaceutical companies. The Commercial Protein Crystallisation Facility was used from May to July 2001, with the unit placed behind a panel in Zvezda, and crystals were returned to Earth on STS-104 along with US crystals from Expedition 2. The original version, CPCF, flew on the John Glenn shuttle flight STS-95.

The Enhanced Gaseous Nitrogen dewar (EGN) had flown on the Mir space station but this enhanced version had a heater to better control the rate at which the nitrogen ice in the dewar bottle melted, allowing crystals to slowly form. There was a problem with bubbles in the sample tubes brought back from the risk mitigation flight but large crystals were still grown. It grew 500 crystals at a time.

Physics of Colloids in Space

Colloids consist of fine particles in suspension in a fluid. Examples are milk, ink and paint. When in gravity, the properties of colloids are masked by buoyancy and sedimentation. Study of colloids in microgravity allows fundamental research and the beginnings of colloidal engineering – making new materials by tailoring colloids. Colloids crystallise in microgravity – they form ordered arrays and lattice structures.

PCS comprises eight 3 ml sample tubes on a carousel for a NASA fluid physics experiment on ISS. Several classes of experiment are conducted:

For binary colloidal alloys the sample comprises two different sizes of particle in a fluid and the carousel is spun up to "homogenise" (evenly distribute) the sample. The sample is then left to form a colloidal super-lattice (alloy). Experiment runs last from 12 to 96 hours. Measurements are made for each sample a few times each week over several months to study "aging" as well as colloid crystal formation.

For another set of experiments polymer is added to a suspension of a single size of particles in a fluid and spun up on the carousel to "homogenise". By using the right amount of polymer you can produce either crystals or gel (a fine structure stretching across the whole sample tube) and get behaviour like a solid, a liquid or a gas. One colloid-polymer crystal sample unexpectedly behaved like a gel in its "aging" (stiffening). The colloid-polymer gel was found to have a resonance near the vibration frequency of mechanical modes of ISS. The colloid-polymer critical point sample separated into regions that behaved like a gas (colloid poor) and a liquid (colloid rich) in a way unlike any ever measured on Earth.

Additional experiments add a salt solution to fine particles in suspension – to produce a fractal gel (like a snowflake, it has structure in ever finer detail as you look at it more closely) – and a sample with so many particles in suspension that it behaves like a glass rather than a crystal. The soot that contaminates diesel lubricating oil is an example of a fractal colloid though in gravity a fractal colloid will fall as sediment before it has chance to form a gel.

The colloids are studied using laser light scattered by the samples and by imaging the samples with a pair of colour cameras (with different magnifications). Data are sent to the ground. The experiment is controlled from the Glenn Research Centre and Harvard University using telescience techniques and largely runs under ground control with occasional maintenance by the crew. The work follows on from glove box work on STS-95 and Mir. The experiment wouldn't activate properly initially and troubleshooting was needed. It was offline in September 2001 due to Express rack computer reboot problems and a computer reload was needed. During an experiment run in September a detector module failed – meaning experiments would take longer and there would be operational changes.

By November 2001, the experiment was going so well that it was pushed to make measurements beyond its original design using the low angle scattering camera and by making time-lapse movies.

Despite computer reboot problems in 2002, which forced early termination of the experiment, NASA considers PCS a significant success. During the 12 months that data was gathered, all 8 sample cells gave interesting new results. In many cases this revealed phenomena that have not been observed in an earthbound laboratory.

Global Transmissions Services [7]

This experiment tests the worldwide reception of an accurate time signal.

GTS comprises an antenna unit and an electronics unit. The AU was launched on Zvezda, fixed to EVA handrails outside the module. The EU was launched on the Progress M45 (5P) freighter in August 2001 (delayed from the early 2001 original launch date) and installed inside Zvezda by the crew in September 2001. An ultra-stable quartz oscillator (USO) inside the EU provides a time signal to users in the Russian segment of ISS. The AU uses phased array antennas to transmit the same time signal on UHF and L-band everywhere along the flight path of ISS. The USO is synchronised to the International Atomic Time (TAI) from the ground. Clocks on Earth can receive the UHF radio signals giving UTC and local time. Prototype desktop clocks are under development right now.

Steinbeis TZR runs the commercial experiment with industrial partners like the watchmaker Fortis. ESA provided the launch opportunity using resources originally planned for use on Euromir-E on Mir in late 1997.

This is an initial demonstration of the concept and an operational system might use additional low Earth orbit satellites to provide a 24 hour service. The initial trials measure signal quality and data rates received on the ground during 5-7 periods of 5-12 minutes per day. The on-orbit checkout via TsUP was on 15-17th January 2002 and the first signal reception at the Stuttgart ground station was on 12th February 2002. What had to be ironed out were interference problems between the UHF transmitter and signals from NASA spacewalker helmet cameras and NASA fears about radiation from the L-band transmitters affecting their spacewalkers. Compare the 120 mW transmitter output to the 2 W from a mobile phone!

Hoffman-Reflex

Studies of humans in space indicated that exercise might become less effective in long duration spaceflight. This is due to nervous system changes in microgravity that cause the spinal cord to be less responsive. The Hoffman reflex (H-reflex) has been a clinical test since the 1800s. A mild electric shock is applied to the back of the knee (it feels like a static electricity shock) and the muscle response is measured by an electromyogram – measuring the electrical activity in a contracting muscle. It's similar to the lower leg jerk when a doctor taps your knee but can be measured more precisely. What's being measured is spinal cord excitability and it's expected to decrease as time in space increases. The effect was noted as a side effect during past investigations but is now being measured directly.

The subject is restrained at waist and feet in a seat like device. The operator (who can be the same person as the subject) can directly control the shock strength. The data goes directly into the HRF laptop and can be displayed directly so that data gathering can be checked.

The experiment is from the Canadian Space Agency (their first aboard ISS). The principal investigator talked with the crew in the course of their flight about the experiment. Measurements are made within the first day in space, after 6 days in space, 12 days before landing and three times after landing (at the 1, 2 and 4 day points). Expedition 3 did an extra set of tests after about 1 month in space at the request of the investigator, based on the results he'd got from Expedition 2.

Early results have shown that the spinal cord reflex does reduce rapidly in microgravity but that it unexpectedly recovers after several months in microgravity.

LSO (Lightning and Sprites Observations) [8]

A French atomic agency (CEA) experiment flown to ISS for use in the Andromède mission by Claudie Haigneré. The pair of digital micro cameras on a swing arm is used to observe lightning and other phenomena. The new classes of object called elves and sprites occur above thunderstorms. They were observed by pilots for many years but not recorded until 1989 – by a low light TV camera being tested before a rocket launch – and are now of major research interest. They fuel the global electrical circuit and may be adding charged particles to Earth's van Allen belts.

Sprites are blobs, columns and plumes that extend from 50-90 km high. They last just 1/10 second but are visible to the naked eye.

Elves are expanding rings up to 400 km wide at a similar altitude to sprites. They are faint and last only 1 millisecond.

Blue jets are much more rare. They are narrow fountains of light shooting up to 50 km high at 100 km/second.

The mechanism causing them is still unknown but all occur in association with large thunderstorm systems – mesoscale convective complexes. They had been detected by sensors on Salyut and as gamma ray bursts by the Gamma Ray Observatory satellite.

LSO is defining parameters for a possible future French microsatellite mission to study lightning phenomena. Astronaut intervention allows identification of areas of high activity that are not always clear on wide-angle cameras.

CBOSS (Cellular Biotechnology Operational Support System) [6, 9]

Microgravity presents a unique opportunity to grow in vitro tissue (outside the body) that resembles in vivo tissue (inside the body). On Earth tissues normally only grow in a 2-dimensional way outside the human body. This means that skin can be grown but it took NASA developed rotating bioreactors to grow small 3-dimensional samples of other tissue. NASA flew such bioreactors in space on shuttle flights from 1995. NASA then developed stationary bioreactors for flight on the Mir space station.

Normally cells grown outside the human body don't differentiate or specialise in the way they would inside the body. They don't self organise on the road to forming tissues or organs and they don't produce the proteins and other chemicals that they would when inside a body. Three-dimensional growth inside a rotating bioreactor on the ground is a step in the right direction and growth in microgravity is a step further. Impressive results were obtained from experiments on Mir and the changes in cell behaviour in microgravity also give insights into microgravity effects on the human body. The cells grown on the ground give rise to small clumps at best but on Mir the whole growth container filled with fluffy tissue.

CBOSS is an extension of this work, comprising the Biotechnology Specimen Temperature Controller (BSTC), a refrigerator (for transport of cells or tissues), a gas supply system (oxygen and carbon dioxide for purging the growth chambers every day to maintain cell metabolic processes) and storage racks. CBOSS is an interim set of equipment, for use until the proposed ISS Biotechnology Facility is ready.

BSTC holds up to 32 TCMs (Tissue Culture Module), which are static tissue culture vessels – cell incubators comprising clear Teflon bags full of growth media maintained at a controlled temperature and atmosphere. The unit has to be kept powered – for a constant temperature of 36 degrees C.

Cell growth starts soon after delivery to ISS. A crewmember monitors its status, adds nutrients (using syringes, gloves and a face mask) and collects samples. Cells and growth media are checked with a portable clinical blood analyser, like the type used in hospitals, for pH and chemical analysis. The samples are preserved for return to Earth by using a syringe to add fixative and placing them in a refrigerator. The samples in the refrigerator were re-arranged by Frank Culbertson due to concern they were getting warmer than the 4 degrees C required.

Tissues grown in space are studied to learn more about the effects of microgravity on the human body, to be used for pharmaceutical research and to model disease behaviour. NASA was originally keen on the idea of whole tissue growth in microgravity but this now seems far in the future and research is at a much more basic level. 24 cell cultures were grown using CBOSS in 2001 [10]:

Ovarian cancers grown in space more closely resemble those from inside the body in their chemotherapy resistance. They will be studied and used as models for the development of new therapies.

Colon cancers were studied to learn what their nutrient consumption was, ready for future bioreactor work.

The practicality of growing neuroendocrine (neural precursor) tissue was studied and they could ultimately find use as transplants in neurological diseases like Parkinson's.

Part of the human kidney was studied to focus on a better understanding of pharmaceutical toxicity.

PuFF

The effects of EVA and long term exposure to microgravity on pulmonary function (PuFF) is a continuation of

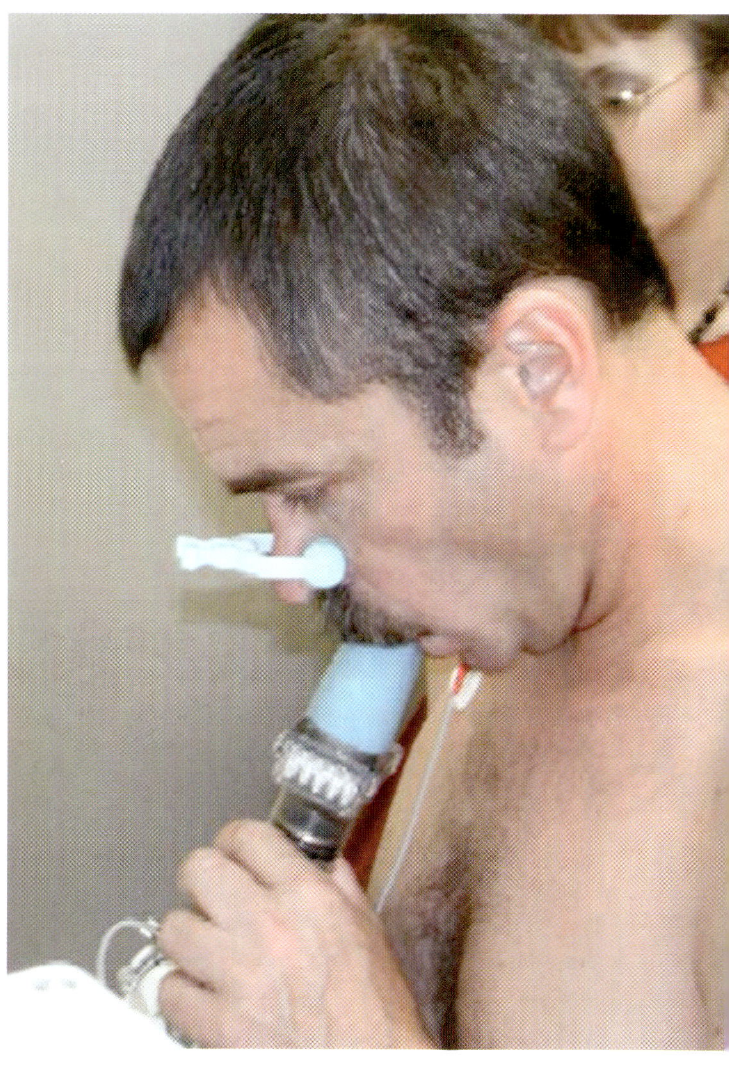

Mikhail Tyurin trains with the Pulmonary Function in Flight experiment hardware prior to his Space Station mission.
(NASA)

spacelab research into how the lungs behave. Just being in microgravity changes them in a manner not researched in detail so far. Since muscles grow weak in microgravity this is one obvious effect. Breathing contaminants and dust particles inside ISS may have an effect on the lungs as well.

To go on spacewalks a person breathes oxygen at reduced pressure (to avoid the spacesuit becoming too rigid to bend). Pre-breathing of pure oxygen is required to purge nitrogen from the bloodstream else it can form bubbles at the lower pressure of a spacesuit. This gives rise to the painful condition called "the bends" – a problem experienced by divers. Even when the bends are not noticeable there may be micro-embolisms present.

PuFF is done 2 weeks into a flight and then monthly. It is also done one week before an EVA and then on the same day after an EVA (or as soon as the schedule allows).

The experiment uses the GASMAP (Gas Analyser System for Metabolic Analysis Physiology) part of the Human Research Facility rack – with the subject breathing cabin air. A PuFF session involves 5 lung function tests to measure gas exchange in the lungs, by studying inhaled and exhaled air, and respiratory muscle strength. Data is stored on the Human Research Facility laptop.

Materials Samples

Three sets of materials samples were attached outside ISS in 2001: MISSE from NASA, Kromka (Edge) from Russia and MPAC&SEED from Japan.

This is the perfect environment for testing materials that have to be exposed for a long duration to the environment of space. The onslaught is from solar radiation (ultra-violet rays), atomic oxygen (from the upper atmosphere), a hard vacuum and contamination by ISS itself.

Susan Helms and laptop computers near the Human Research Facility.　　　(NASA)

MISSE is a collaborative effort from NASA, Boeing and the US Air Force. The two suitcase sized Passive Experiment Containers were originally deployed outside the Mir space station by NASA but were retrieved in 1998.

More than 700 samples include thermal coatings, cosmic ray shields, inflatable structure membranes, solar sail material, optics, paint, advanced solar cells and tether materials. The samples were selected based on projected future needs for civil and military missions. The PECs were attached by spacewalkers outside the US Quest airlock in August 2001 and will be exposed to space for about one year. It was the first external experiment to be deployed from ISS. Some of its parameters are monitored from the ground. Researchers were interested in a deployment during solar maximum – to compare with that on Mir near solar minimum.

Kromka (Edge) was deployed close to thruster nozzles on the Zvezda module – both to collect nozzle exhaust particles and to test protective coatings on materials.

The sample cassettes will be retrieved and replaced by spacewalkers over time – like similar experiments outside the Mir space station. Two of the samples included in an EVA deployed unit called Epsilon were Zerodur mirror surfaces (used on X-ray telescopes) and another two were of thin CCD crystal surfaces. An extra sample from the University of Leicester was of Micro Channel Plate (a detector) to be used in a possible future ISS X-ray telescope (LOBSTER).

MPAC&SEED (Microparticles capture and Space Environment Exposure Device) was studying both materials samples and micrometeoroid levels for Japan. Three double sets of panels were deployed outside the Zvezda module by spacewalkers in October 2001.

One set will be retrieved by EVA every year for 3 years and returned to Earth inside Soyuz. These will measure both space debris and natural micrometeoroid levels after analysis.

The SEED samples include solid lubricants, insulation, adhesives, paints and thin films for use in inflatable structures. MPAC used Aerogel and polyamide foams as capture media.

Conclusion

The goals of research on ISS are best summed up by NASA OBPR associate administrator Mary Kicz's testimony to congress in 2002:

"It is about having the capability to perform multiple repetitions of experiments when they hint at new discoveries, and the ability to modify and re-fly the most promising of these experiments. It's about studying multiple generations of

Mikhail Tyurin and the MPAC-SEED experiment before EVA deployment. (NASA)

several types of organisms in space to learn about development, growth and aging in the absence of gravity. It's about using finely controlled experiment environments to test fundamental scientific theories with a precision that is unparalleled on the ground".

Acknowledgments

Very special thanks to Increment Lead Scientist Dr John Uri and to Prof Peter Pusey (University of Edinburgh), Roberta Castagini, Dr Stefan Hofer (Kayser-Threde), Dr Juergen Stapelmann (Astrium) and Dr Felix Huber (GTS)

References

1. See p.33 Expedition 1 Press kit (NASA) for the original Russian experiments list.
2. *Physical Review Letters*, **83**, No. 8, p..1598 The American Physical Society.
3. Radiation and the International Space Station: recommendations to reduce risk, National Academy Press (National Research Council).
4. Eol.jsc.nasa.gov.
5. Eos Transactions, American Geophysical Union 23/04/02.
6. Future Biotechnology Research on the International Space Station, National Academy Press (National Research Council)
7. On Station No.5, ESA, March 2001, *Spaceflight*, **42**, January 2000.
8. Big sky, hot nights, red sprites, Scientific American presents Weather, 11, No.1, Spring 2000.
9. A biorevolution through tissue engineering Chapter 4 (P.81-118), Challenges of human space exploration, M. Freeman, Springer-Praxis, 2000.
10. Evaluation of Ovarian Tumor Cell Growth and Gene Expression: Jeanne Becker from the University of South Florida, Mechanisms of Colon Carcinoma Metastasis in Microgravity: J. Millburn Jessup from the University of Texas Health Science Centre, PC12 Pheochromocytoma Cells: Peter I Lelkes of Drexel University, Human Renal Cortical Cell Differentiation and Hormone Production: Timothy Hammond of Tullane University Medical Centre.

Other References

Web sites:

Space research in Destiny
http://spaceresearch.nasa.gov/
RKK Energiya (ISS)
http://www.energia.ru/english/energia/iss/
ISS science operations news (MSFC)
http://www.scipoc.msfc.nasa.gov/
Microgravity Research Program Office (NASA HQ)
http://mrpo.nasa.gov/iss/
NASA Human spaceflight (ISS)
http://spaceflight.nasa.gov/station/
Expedition and Shuttle press kits
http://www.shuttlepresskit.com

ISS User's Guide Release 2.0.
Aviation Week & Space Technology, 8 December 1997, p.42+ (ISS special issue).
Aviation Week & Space technology July 9, 2001 P. 57 ("Limits on ISS lab use slow science schedule").
ISSCOMM reports by Chris van den Berg.
The International Space Station, A guide for European users. ESA BR-137. February 1999.
Dr Erik Seedhouse, "Radiation concerns for astronauts", *Spaceflight*, **41**, pp.323-325, 2002.

To follow activities on ISS read "Space Station Chronology" by Neville Kidger in *Spaceflight* each month.

ISS Soyuz Crewing 2000-2001

BERT VIS

Introduction

Given the limited lifetime of the Soyuz, which was to be used as ISS ferry craft and essentially as the life-boat for the expedition crews, it had to be exchanged every half year for a new one. The profile for these replacement missions was a rather straightforward one: a crew would deliver a new Soyuz to the station, stay there for a few days and return to Earth on the "old" ship.

This had also been the practice during the Mir programme, although there it was not always a visiting crew that brought up the new ship. In particular near the end of the Mir programme, it was the new expedition crew that would fly to the station with a new Soyuz after which the old crew would return with their own ship. But since beginning with Expedition 2 every crew would be transported to and from ISS with the space shuttle, a separate crew would take care of refreshing the ferry craft. This would soon lead to the nickname that these missions got: Soyuz Taxi flights.

SOYUZ TM-31
Prime Crew: Yuri P. Gidzenko, Sergey K. Krikalev, William M. Shepherd
Backup Crew: Vladimir N. Dezhurov, Mikhail V. Tyurin, Kenneth D. Bowersox

When Soyuz TM-31 lifted off from Baykonur on 31 October 2000, the three cosmonauts it carried in all probability comprised the best-trained crew ever to fly in space. The first crew to live and work on the space station had been assigned for the mission on 30 January 1996 and therefore had spent almost five years preparing for the flight. On top of this, all three had previously flown in space: while Bill Shepherd had logged a little over 18 days in space on three space shuttle missions [1], the Soyuz commander, Yuri Gidzenko, had made one flight to the Mir space station that had lasted 179 days. The flight-engineer, Sergey Krikalev, was by far the most experienced crewmember, with four flights and a total of 484 days logged in orbit [2].

The sheer fact that the Russians had so much more experience than their American counterparts had actually been the reason for the first crew change. Initially, the Soyuz would have been commanded by cosmonaut Anatoliy

Expedition One prime and backup crew members in Moscow (from left): Mikhail V. Tyurin, Kenneth D. Bowersox, Vladimir N. Dezhurov, William M. Shepherd, Sergei K. Krikalev and Yuri P. Gidzenko.   (NASA)

Solovyov. But the fact that it had been decided that Shepherd would take over command as soon as the crew would have entered the station, had made Solovyov decide to step down shortly after being assigned. With his vast experience [3], Solovyov's pride didn't allow him to hand over command of the mission to Shepherd, whose experience he considered to be only a fraction of his own, and in October 1996 he stepped down, his place being taken by Gidzenko.

SOYUZ TM-32

Prime Crew: Talgat A. Musabayev, Yuri M. Baturin, Dennis A. Tito
Backup Crew: Viktor M. Afanasyev, Konstantin M. Kozeyev

Soyuz TM-32 was scheduled to be the first Soyuz replacement flight to ISS and would become better known as Taxi-1. This was a bit peculiar, as it would have been more correct to give that name to TM-31, which after all had been the first ferry craft to fly to the station, even though the crew had been the first expedition crew.

Of course, the flight profile for these missions was a good opportunity to give unflown cosmonauts their first spaceflight experience. For decades, it had been a rule in the Russian manned space programme that every crew had to have at least one experienced crewmember. In July 1997, the first crews were assigned. Commander would be Talgat Musabayev with Nadezhda Kuzhelnaya as flight-engineer [4]. Backups would be Vladimir Dezhurov and Sergey Revin. Three months later however, Dezhurov was reassigned as backup for the first ISS long-duration crew and replaced by Valeriy Tokarev. It was the first in a series of crew changes.

The next one occurred in August 2000, when Sergey Revin lost his position as backup flight engineer to Konstantin Kozeyev.

Revin had been informed by RKK Energiya's chief cos-monaut, Gennadiy Strekalov, of the fact that he was re-placed, but was not given a reason. In the Gagarin Cosmo-naut Training Centre however, word had it that he was considered too slow a student, in particular when it came to his English. On the other hand, at Energiya, Revin had worked on the Soyuz TMA spacecraft and now that it was clear that the first two Taxi missions would be flown with the old TM, it may have been that the Energiya brass wanted to move him to that flight. Whatever the case, when Revin was informed of his removal from the backup crew, Strekalov had assured him there was no reason to worry and that he would get his chance to fly. But also Musabayev-Kuzhelnaya and Tokarev-Kozeyev would not be the final crews for Soyuz TM-32.

The Mir space station had still been in orbit when Ameri-can businessman Dennis Tito had expressed interest to fly in space as the first private citizen to do so at his own expense. The first reports that a deal was being negoti-ated became public only three days after Sergey Zalyotin and Aleksandr Kaleri had returned to Earth. The next flight to Mir was to be conducted by Salizhan Sharipov and Pavel Vinogradov. Backups would be Talgat Musabayev and Yuri Baturin, meaning that Musabayev was relieved of his assignment as commander of the first Taxi flight. In September 2000, he was replaced by Viktor Afanasyev.

But while time progressed, it became increasingly clear that the mission would be a close-out one to prepare Mir for it's deorbit. It would certainly not be a flight to take along a tourist, if the mission would get off the ground to

Soyuz TM-32 prime crew. (From left to right): Dennis A. Tito, Talgat A. Musabayev and Yuri M. Baturin,

Soyuz TM-32 backup crew. (From left to right): Viktor M. Afanasyev and Konstantin M. Kozeyev.

begin with. As it was clear that Mir was nearing the end of its lifetime, the Russian negotiators offered Tito a flight to ISS instead of Mir. Tito happily agreed.

With this, a new problem arose however. Tito, who was known among trainers as "Louis de Funes" [5], refused to learn Russian which, incidentally, didn't do much for his popularity in Star City. It meant that the Russians had to come up with a solution for what would be a potential problem if Tito would fly. However, this problem was quickly solved when Musabayev and Baturin were teamed up again, and assigned as prime crew together with Tito for the first Soyuz swap mission. As a result Afanasyev was moved back into the backup commander's slot. Kuzhelnaya and Tokarev lost their assignments altogether, leaving Kozeyev as Afanasyev's flight engineer.

While the crews went through their training for the mission, a political struggle arose between the Russians and NASA. NASA was dead against non-professional astronauts flying to the station. The reason they gave was that the station was still under construction and that the presence of what essentially was a tourist was endangering the mission of the expedition crew and was a safety hazard to ISS as a whole. Most independent observers thought this was nonsense and just an excuse to try and keep Tito off ISS.

The Russians didn't bow to the American protests, stating that it was a Russian decision who would be on their crews and that NASA had no veto over that, just like NASA would never accept a Russian veto over who would fly to the station on the shuttle.

As the mission got closer to launch day, the argument got more grim. In particular NASA administrator Dan Goldin was opposed to Tito flying and did everything he could to make that clear to the Russians. On 12 April, when the 40th anniversary of Gagarin's flight was celebrated, not a single NASA representative attended the festivities. Several trainers made it clear to other foreigners who were present that this was considered a downright insult to the Russians.

Dennis Tito

(Bert Vis)

With NASA not showing up, Dennis Tito was the only American present, and when a group photo was made on the steps of the central building in Star City, he stood next to Russian president Vladimir Putin which was a subtle indication that Putin approved of Tito flying two weeks later. The next day, during another official event, the upcoming crew was introduced to the public, and while Musabayev and Baturin got a good round of applause, Tito got a downright ovation when his name was called.

The argument over Tito flying or not was embarrassing to the astronauts. During a fireworks display in Star City, foreigners overheard two of them talking about the subject and it was clear that the astronauts didn't care one bit who was going to be on that Soyuz, as long as it would come, ensuring that they could continue their own mission.

On 25 April, Soyuz TM-32 was launched making Tito the first private citizen in space paying for his own trip. NASA was still not happy with it though and would not allow Tito to enter the American segment of the station. Also, Goldin ordered his astronauts not to make any photos of Tito in orbit with NASA cameras.

When the flight came to an end, and the crews said goodbye to each other, video showed how Tito wanted to hug Jim Voss who refused, shrugging as if he wanted to say: "I can't help it. Orders...." It didn't seem to bother Tito too much, especially since off-camera there had not been any animosity. Whatever the case, Tito had set a goal and had succeeded in reaching it. To him, that was the main thing.

SOYUZ TM-33

Prime Crew: Viktor M. A Afanasyev, Konstantin M. Kozeyev, Claudie Haigneré
Backup Crew: Sergey V. Zalyotin, Nadezhda V. Kuzhelnaya

Unlike on the previous mission, the crew of the second Soyuz replacement mission was not controversial. This time, professional cosmonauts made up the crew: while the Soyuz TM-32 backups, Viktor Afanasyev and Konstantin Kozeyev, would move on to become the prime crew, in December 2002 it was announced by a public affairs official that Claudie Haigneré had been assigned to fly on TM-33 [6].

When asked by one of the reporters present who would be her backup, Haigneré said that no decision had been made yet but that the way things looked, there might not be a backup at all. In the end this was what happened, the first time that ESA had not assigned a backup for a mission with the Russians.

The 10-day flight was launched from Baykonur on 21 October 2001. Haigneré who had been trained as Flight Engineer, held the designation FE 1, while Kozeyev was FE 2. During her stay onboard ISS, Haigneré and her colleagues conducted an experiment programme that had been set up by the French space agency CNES.

Soyuz TM-33 prime crew. (From left to right):
Konstantin M. Kozeyev, Viktor M. A Afanasyev
and Claudie Haigneré.

Soyuz TM-33 backup crew. (From Right to Left): Nadezhda V. Kuzhelnaya and Sergey V. Zalyotin.

Conclusion

By the time Haigneré flew, the Soyuz replacement missions were generally referred to as "Soyuz Taxi" flights and received more publicity than the expedition crews. The main reason for this was the fact that it had become clear that the Russians were willing to sell the third seat of every flight to whoever would come up with the money for it, generally thought to be around 20 million dollars. A candidate for the next mission was already training in Star City. It was South African businessman Mark Shuttleworth who, like Tito before him, was on the way to realising a life-long dream.

With two such flights taking place every year for the foreseeable future there are ample opportunities for well-to-do people and for organisations to secure a place on board one of them. Given the many press reports on the subject, there seems to be more than enough interest. In many cases however, the financial part of it all remains the problem....

References

1. STS-27, STS-41 and STS-52.
2. Mir EO-4, Mir EO-9/10, STS-60 and STS-88.
3. Shepherd's 18 days, not even as a commander, and total lack of EVA experience could not stand in the shadow of Solovyov's 651 days in orbit on Soyuz TM-5, Mir EO-5, Mir EO-12, Mir EO-19 and Mir EO-24, and record 16 EVAs totalling over 82 hours outside the Mir station.
4. Kuzhelnaya would have been only the fourth Russian woman to fly in space, after Valentina Tereshkova, Svetlana Savitskaya and Yelena Kondakova.
5. Louis de Funes was a famous French film comic; Tito looked a little like him.
6. The announcement took place during photo opportunity at the ESA establishment in the Netherlands, ESTEC, on 15 December 2002. Remarkably, ESA never issued a press release announcing Haigneré's assignment.

*   *   *

The International Space Station -
Orbital Considerations and Related Topics

PHILLIP S. CLARK

Introduction

This review is built around two major tables. The first one is a listing of all of the orbital manoeuvres completed by the International Space Station (ISS) or visiting spacecraft while docked from the time that the first module was launched in 1998 through to the end of 2001. The second is a listing of all of the spacecraft - manned and unmanned, Russian and American - which have been launched towards ISS on docking missions.

Naming the International Space Station

The International Space Station does not have a "catchy, official" name, and some NASA sources have said that such a name would not be appropriate until the station's assembly has been completed. The popular name is simply "ISS", but in some quarters the name "Alpha" has been used.

The history of the name "Alpha" is that when NASA was re-thinking its space station options after it became clear that the mid-1980s Freedom was too large a project to fly, three design options were considered: A, B and C. The combination of options A and B mutated to become Design Alpha, and this evolved into the International Space Station. When the first resident crew boarded the station in November 2000 they adopted the callsign "Alpha", and some elements of the media took this as representing the name of the station.

Thinking back to the Russian experience in regularly flying crews to Salyut and then Mir orbital stations, each mission commander chose his own call sign, and the other crew members adopted this. For example, the first Salyut [1] received the Soyuz 11 crew, and the mission commander Georgi Dobrovolsky adopted the callsign *Yantar*: the three crewmembers used callsigns *Yantar*-1, *Yantar*-2 and *Yantar*-3. While this happened, no-one ever seriously suggested that Salyut had been renamed *Yantar*.

One should therefore look upon the adoption of the "Alpha" callsign by the ISS-1 resident crew in the same way as the Soyuz 11/*Yantar* use *vs* the correct name of the station being Salyut.

If one wanted to play Devil's Advocate, one could always adopt the philosophy that the first modular space station – the Mir Complex – took its name from the designator of the base module, the first element to be launched. Similarly, ISS could become known as the "Zarya Complex" ...

It is worth noting that in April 2001 the Russians quietly celebrated a little-publicised anniversary. On April 22 the Russians celebrated the 30th anniversary of the last time that they had to cancel a piloted launch with the crew in the spacecraft (Soyuz 10, the first crew to visit Salyut). While there have been missions which have been delayed because of financial considerations, plus the launch failures in April 1975 and September 1983, this record is remarkable – especially when compared with the regular media event of US crews regularly making return trips from un-launched shuttles.

Comments Regarding the Main Tabular Listings

All of the orbital data used in this paper are derived from the Two-Line Orbital Elements (TLEs) which are generated with United States Space Command (USSPACECOM) and issued *via* the World-Wide Web site operated by the Orbit Information Group at the NASA Goddard Space Flight Center [2]. Software to convert from the TLE format to more recognisable orbital parameters has been written by this author, using constants which are consistent with those used within USSPACECOM for its orbital modelling.

Table 1 is a master listing of all of the orbital manoeuvres conducted by spacecraft which are either permanent modules of the ISS assembly or visiting spacecraft – either US shuttles or Russian Progress-M/M1 cargo freighters. Prior to the launch of the second Russian module, ISS/Zvezda, the "base module", ISS/Zarya, completed orbital manoeuvres, but the arrival of ISS/Zvezda meant that any further manoeuvres by ISS/Zarya would involve the danger of damaging the new module with propellant deposits.

As well as listing all of the orbital manoeuvres conducted by ISS, Table 1 also lists the orbit of the station immediately after the docking (or re-docking in a few cases) of spacecraft. The final column shows the ISS element (actually ISS/Zarya before the arrival of ISS/Zvezda) or visiting spacecraft which performed each manoeuvre. As ISS has become more massive, the visiting Progress-M/M1 spacecraft have become less effective in manoeuvring the assembly, and therefore since ~September 2000 the main orbit maintenance manoeuvres have been performed by visiting shuttle orbiters.

TABLE 1: *List of Manoeuvres of ISS, 1998-2001.*

Pre-Manoeuvre Orbit						Post-Manoeuvre Orbit						Manoeuvring Spacecraft
Orbital Epoch	Orbital Inclination °	Orbital Period min	Perigee km	Apogee km	Arg of Perigee °	Orbital Epoch	Orbital Inclination °	Orbital Period min	Perigee km	Apogee km	Arg of Perigee °	
1998						**1998**						
Initial orbit of ISS/Zarya						Nov 20.28	51.59	89.72	177	344	86	
Nov 21.02	51.59	89.68	177	341	90	Nov 21.45	51.59	90.44	246	345	97	ISS/Zarya
Nov 23.15	51.60	90.41	245	344	105	Nov 23.35	51.59	91.47	309	384	53	ISS/Zarya
Nov 23.35	51.59	91.47	309	384	54	Nov 23.46	51.60	91.48	311	384	59	ISS/Zarya
Nov 24.10	51.59	91.48	309	385	63	Nov 24.48	51.58	92.38	385	397	181	ISS/Zarya
Docking of Endeavour/STS-88 + ISS/Unity						Dec 6.15	51.60	92.33	383	395	263	
Dec 8.71	51.60	92.31	382	394	275	Dec 9.00	51.60	92.44	388	401	227	Endeavour/STS-88
Dec 16.73	51.60	92.42	387	399	263	Dec 16.92	51.58	92.50	388	406	251	ISS/Zarya
Dec 21.48	51.59	92.49	394	399	283	Dec 21.99	51.60	92.56	397	403	296	ISS/Zarya
1999						**1999**						
						Jan 1.01	51.59	92.54	396	402	331	
Docking of Discovery/STS-96						May 29.49	51.59	92.21	381	386	162	
Jun 3.42	51.59	92.20	379	386	215	Jun 3.61	51.60	92.40	385	400	299	Discovery/STS-96
Oct 26.19	51.59	92.00	366	379	120	Oct 26.76	51.59	92.03	366	383	115	ISS/Zarya
Dec 1.88	51.59	91.84	358	372	263	Dec 2.04	51.59	92.20	377	388	270	ISS/Zarya
2000						**2000**						
						Jan 1.10	51.59	92.09	372	383	28	
Docking of Atlantis/STS-101						May 21.64	51.58	91.17	328	336	231	
May 23.91	51.58	91.15	327	335	240	May 24.09	51.58	91.47	342	352	357	Atlantis/STS-101
May 24.93	51.58	91.46	339	353	332	May 25.18	51.58	91.78	356	367	260	Atlantis/STS-101
May 25.65	51.58	91.77	356	367	259	May 26.20	51.58	92.08	372	381	347	Atlantis/STS-101
Jul 18.17	51.58	91.88	363	370	183	Jul 18.31	51.59	91.95	367	373	146	ISS/Zarya
Jul 25.92	51.58	91.89	364	371	167	Jul 26.22	51.58	91.74	351	368	172	ISS/Zarya
Docking of ISS/Zvezda						Jul 28.26	51.58	91.72	351	367	178	
Docking of Progress-M1 3						Aug 8.83	51.58	91.67	349	364	224	
Aug 15.40	51.58	91.64	348	363	247	Aug 15.84	51.58	91.68	350	363	272	Progress-M1 3
Aug 17.54	51.58	91.67	349	364	246	Aug 17.93	51.58	91.82	357	370	314	Progress-M1 3
Docking of Atlantis/STS-106						Sep 10.49	51.58	91.73	353	366	40	
Sep 11.45	51.58	91.72	353	365	44	Sep 11.71	51.58	91.85	360	370	39	Atlantis/STS-106
Sep 14.08	51.58	91.83	361	368	52	Sep 14.45	51.58	91.96	369	371	102	Atlantis/STS-106
Sep 15.21	51.58	91.95	368	372	126	Sep 15.40	51.58	92.09	373	381	118	Atlantis/STS-106
Sep 17.11	51.58	92.07	373	379	189	Sep 17.38	51.58	92.19	377	387	223	Atlantis/STS-106
Docking of Discovery/STS-92						Oct 13.87	51.57	92.12	375	382	327	
Oct 16.37	51.58	92.11	374	381	344	Oct 17.21	51.58	92.16	379	381	339	Discovery/STS-92
Oct 17.82	51.57	92.16	380	381	328	Oct 18.16	51.58	92.22	381	386	62	Discovery/STS-92
Oct 18.84	51.57	92.22	381	386	56	Oct 19.25	51.58	92.28	382	391	82	Discovery/STS-92
Docking of Soyuz-TM 31 **First Resident Crew**						Nov 2.63	51.58	92.22	380	387	147	
Docking of Progress-M1 4						Nov 18.19	51.57	92.16	377	384	197	
Docking of Endeavour/STS-97						Dec 3.20	51.57	92.08	374	379	275	
Re-docking of Progress-M1 4						Dec 26.57	51.58	91.94	365	375	273	
2001						**2001**						
						Jan 1.14	51.58	91.90	362	373	293	
Docking of Atlantis/STS-98 + ISS/Destiny						Feb 9.88	51.57	91.67	350	363	87	

TABLE 1: *List of Manoeuvres of ISS, 1998-2001 (Contd).*

Pre-Manoeuvre Orbit						Post-Manoeuvre Orbit						Manoeuvring Spacecraft
Orbital Epoch	Orbital Inclination °	Orbital Period min	Perigee km	Apogee km	Arg of Perigee °	Orbital Epoch	Orbital Inclination °	Orbital Period min	Perigee km	Apogee km	Arg of Perigee °	
Feb 11.54	51.58	91.68	351	363	88	Feb 11.85	51.57	91.75	351	370	31	Atlantis/STS-98
Feb 11.91	51.56	91.75	356	365	86	Feb 12.04	51.57	91.81	356	371	94	Atlantis/STS-98
Feb 13.76	51.57	91.79	355	370	109	Feb 13.96	51.57	91.99	366	378	98	Atlantis/STS-98
Feb 14.91	51.58	91.98	365	378	100	Feb 15.18	51.58	92.03	367	381	113	Atlantis/STS-98
Feb 15.67	51.57	92.03	367	381	109	Feb 16.14	51.58	92.20	376	389	104	Atlantis/STS-98
Re-Docking of Soyuz-TM 31						Feb 24.56	51.57	92.17	374	387	135	
Docking of Progress-M 44						Feb 28.47	51.58	92.14	373	385	151	
Docking of Discovery/STS-102						Mar 10.37	51.57	92.06	370	381	188	
Mar 14.53	51.57	92.04	369	380	206	Mar 14.68	51.57	92.17	372	389	187	Discovery/STS-102
Mar 16.52	51.57	92.16	372	389	196	Mar 16.65	51.57	92.23	377	391	192	Discovery/STS-102
Mar 17.22	51.57	92.23	377	391	195	Mar 17.54	51.57	92.31	381	394	185	Discovery/STS-102
Apr 8.44	51.57	92.15	376	384	253	Apr 9.83	51.55	92.29	379	395	263	Progress-M 44
Re-docking of Soyuz-TM 31						Apr 18.80	51.57	92.23	376	393	294	
Docking of Endeavour/STS-100						Apr 22.50	51.57	92.21	374	391	305	
Apr 23.36	51.57	92.20	374	391	308	Apr 24.00	51.57	92.27	377	395	324	Endeavour/ STS-100
Apr 27.13	51.5	792.25	377	393	333	Apr 27.61	51.57	92.42	382	405	346	Endeavour/ STS-100
Docking of Soyuz-TM 32						Apr 30.84	51.57	92.40	381	403	358	
Docking of Progress-M1 6						May 23.34	51.57	92.29	375	399	76	
Docking of Atlantis/STS-104 + ISS/Quest						Jul 14.21	51.57	92.05	364	386	283	
Jul 15.95	51.57	92.04	363	386	290	Jul 16.19	51.58	92.11	367	389	284	Atlantis/STS-104
Jul 18.29	51.57	92.10	367	388	293	Jul 18.76	51.57	92.18	372	390	290	Atlantis/STS-104
Jul 19.32	51.57	92.17	372	390	292	Jul 19.45	51.57	92.34	381	397	270	Atlantis/STS-104
Jul 22.89	51.57	92.33	380	397	286	Jul 23.16	51.59	92.37	382	399	296	Progress-M1 6
Jul 23.75	51.58	92.36	382	399	299	Jul 24.06	51.60	92.39	382	401	304	Progress-M1 6
Jul 24.88	51.60	92.39	382	401	310	Jul 25.14	51.62	92.43	385	403	302	Progress-M1 6
Jul 25.91	51.62	92.42	385	402	307	Jul 26.12	51.64	92.45	386	403	302	Progress-M1 6
Docking of Discovery/STS-105						Aug 12.81	51.64	92.40	385	400	7	
Aug 14.60	51.64	92.39	384	399	14	Aug 15.00	51.63	92.46	389	402	10	Discovery/ STS-105
Aug 17.49	51.64	92.42	387	400	18	Aug 17.79	51.64	92.50	392	402	12	Discovery/ STS-105
Docking of Progress-M 45						Aug 23.39	51.64	92.48	391	402	26	
Docking of ISS/Pirs						Sep 17.10	51.64	92.36	385	396	105	
Sep 30.88	51.64	92.27	381	391	148	Oct 18.83	51.64	92.47	390	401	216	Progress-M 45
Re-docking of Soyuz-TM 32						Oct 19.51	51.64	92.47	390	401	219	
Docking of Soyuz-TM 33						Oct 23.88	51.64	92.42	388	399	235	
Docking of Progress-M1 7						Nov 28.95	51.64	92.16	373	387	30	
Docking of Endeavour/STS-108						Dec 7.90	51.64	92.09	370	384	69	
Dec 9.75	51.64	92.09	372	382	135	Dec 9.87	51.64	92.15	375	385	60	Endeavour/ STS-108
Dec 11.73	51.64	92.13	374	384	81	Dec 11.85	51.64	92.20	379	386	65	Endeavour/ STS-108
Dec 12.69	51.64	92.19	378	386	80	Dec 12.89	51.64	92.34	388	390	17	Endeavour/ STS-108
Dec 31.92	51.64	92.21	381	385	81	Last set of orbital data for period under review						

Notes: The orbital data in this Table are derived from the Two-Line Orbital Elements which are issued *via* the NASA Goddard Space Flight Center's Orbit Information Group (OIG) web site. The first set of orbital data following the docking (or re-docking) of spacecraft is shown above, as well as the final pre-manoeuvre and first post-manoeuvre set of data. For convenience, the first set of orbital data issued each calendar year is also shown. For manoeuvres the spacecraft which performed the manoeuvre is listed. "Arg of Perigee" is the argument of perigee.

Table 2 is a master listing of all of the spacecraft which have been launched by the Russians and the United States in the ISS programme. Launch, docking, undocking, (re-docking when appropriate,) and landing/de-orbit dates and times are shown, as well as the initial orbit of each spacecraft and the orbit used when docking (re-docking) with ISS. The port used by the spacecraft for each docking is shown.

TABLE 2: *List of Spacecraft Docked to ISS, 1998 to 2001.*

Spacecraft	Launch Date/Time	Docking Port		Docking Date/Time	Initial Orbit				Epoc
					Epoch	Incl o	Period min	Altitude km	
ISS/Zarya	1998 Nov 20 06.40				Nov 20.28	51.59	89.72	177-344	
Endeavour/STS-88	1998 Dec 4 08.36	Un/PMA-1	+YL	1998 Dec 6 02.48	Dec 4.38	51.59	89.54	181-323	Dec 6
+ ISS/Unity		Za	+YL						
Discovery/STS-96	1999 May 27 10.50	Un/PMA-2	+YL	1999 May 29 04.24	May 27.44	51.59	91.19	324-342	May 29
Atlantis/STS-101	2000 May 19 10.11	Un/PMA-2	+YL	2000 May 21 04.31	May 19.45	51.58	89.39	159-329	May 21
ISS/Zvezda	2000 Jul 12 04.56	Za	-YL	2000 Jul 28 00.45	Jul 12.38	51.60	89.63	179-334	Jul 28
Progress-M1 3	2000 Aug 6 18.27	Zv	-YL	2000 Aug 8 20.14	Aug 6.94	51.58	88.56	185-222	Aug 9
Atlantis/STS-106	2000 Sep 8 12.46	Un/PMA-2	+YL	2000 Sep 10 05.51	Sep 8.58	51.58	89.34	159-325	Sep 11
Discovery/STS-92	2000 Oct 11 23.17	Un/PMA-2	+YL	2000 Oct 13 17.45	Oct 12.00	51.57	89.32	159-322	Oct 13
Soyuz-TM 31	2000 Oct 31 07.53	Zv	-YL	2000 Nov 2 09.24	Oct 31.38	51.66	88.65	185-231	Nov 2
		Za	-ZN	2001 Feb 24 10.37					Feb 24
		Za	-YL	2001 Apr 18 13.01					Apr 18
Progress-M1 4	2000 Nov 16 01.33	Za	-ZN	2000 Nov 18 03.07	Nov 16.24	51.64	88.53	186-217	Nov 18
		Za	-ZN	2000 Dec 26 10.54					Dec 26
Endeavour/STS-97	2000 Dec 1 03.06	Un/PMA-3	-ZN	2000 Dec 2 19.59	Dec 1.16	51.58	89.71	197-323	Dec 2
Atlantis/STS-98	2001 Feb 7 23.13	Un/PMA-3	-ZN	2001 Feb 9 16.51	Feb 7.99	51.57	89.78	205-322	Feb 9
+ ISS/Destiny		Un	+YL	2001 Feb 10 19.00					Feb 28
Progress-M 44	2001 Feb 26 08.09	Zv	-YL	2001 Feb 28 09.50	Feb 26.39	51.57	88.64	191-224	
Discovery/STS-102	2001 Mar 8 11.42	De/PMA-2	+YL	2001 Mar 10 06.39	Mar 8.54	51.56	88.44	159-235	Mar 10
Endeavour/STS-100	2001 Apr 19 18.41	De/PMA-2	+YL	2001 Apr 21 13.59	Apr 19.80	51.57	89.39	159-329	Apr 21
Soyuz-TM 32	2001 Apr 28 07.37	Za	-ZN	2001 Apr 30 07.58	Apr 28.37	51.57	88.62	182-231	Apr 30
		Pirs	-ZN	2001 Oct 19 11.07					Oct 19
Progress-M1 6	2001 May 20 22.33	Zv	-YL	2001 May 23 00.24	May 20.99	51.58	88.56	189-218	May 23
Atlantis/STS-104	2001 Jul 12 09.08	De/PMA-2	+YL	2001 Jul 14 03.08	Jul 12.40	51.57	88.41	156-236	Jul 14
+ ISS/Quest		Un	+X	2001 Jul 15 07.34					
Discovery/STS-105	2001 Aug 10 21.10	De/PMA-2	+YL	2001 Aug 12 18.42	Aug 11.04	51.64	89.27	199-278	Aug 12
Progress-M 45	2001 Aug 21 09.24	Zv	-YL	2001 Aug 23 09.51	Aug 21.44	51.64	88.64	185-229	Aug 23
Pirs module	2001 Sep 14 23.35				Sep 15.03	51.63	88.53	190-214	Sep 17
+ ISS/Pirs		Zv	-ZN	2001 Sep 17 01.05					
Soyuz-TM 33	2001 Oct 21 08.59	Za	-ZN	2001 Oct 23 10.44	Oct 21.67	51.64	88.26	182-195	Oct 23
Progress-M1 7	2001 Nov 26 18.24	Zv	-YL	2001 Nov 28 19.43	Nov 26.76	51.61	88.66	190-227	
Endeavour/STS-108	2001 Dec 5 22.19	De/PMA-2	+YL	2001 Dec 7 20.03	Dec 5.96	51.64	89.05	220-235	Dec 7

Notes: Dates and times of events are shown as hh.mm and are based upon GMT: orbital epochs are shown as decimals of a day, GMT. Orbital data sh due to minor inaccuracies in the Two-Line Orbital Elements. The years of the orbital epochs are not shown, but they are self-evident when the orbits are

Permanent modules of the ISS Complex are prefixed by ISS.

 * Date/time of shuttle orbiter undocking from ISS after leaving a new module attached to the station
 ** Date/time (when latter available) of Progress de-orbit
 *** Pirs (instrument) module was docked at the rear of Pirs and was separated after Pirs after it had been attached to ISS

Modules to which spacecraft docked are: De - ISS/Destiny, Pirs - ISS/Pirs, Un - ISS/Unity, Za - ISS/Zarya, Zv - ISS/Zvezda. The docking ports on each longitudinal (long) axis and N along the nadir axis. PMA is an abbreviation for "Pressurised Mating Adapter".

The docking orbits shown in Tables 1 and 2 can differ slightly. This is because USSPACECOM issues separate sets of Two-Line Orbital Elements for ISS itself (used in Table 1) and the visiting spacecraft (Table 2), and very slight differences in the derived TLEs can result in the orbital altitude of a docked spacecraft differing by ~1-2km compared with the station itself.

At the end of 2001 there were two visiting spacecraft attached to ISS: the emergency-return Soyuz-TM 33 piloted spacecraft and the Progress-M1 7 cargo freighter.

Cataloguing Objects in Orbit

The assembly of modular orbital stations have resulted in an inconsistent policy being adopted by USSPACECOM, dating back to the Mir Complex and continuing through with the evolution of the ISS Complex. The discussion here will highlight these inconsistencies.

cl	Period min	Altitude km	Undocking Date/Time	Descent Date/Time
				Permanent Component of ISS
59	90.56	214-390	*1998 Dec 13 20.30	*1998 Dec 16 03.53
				Permanent Component of ISS
59	92.21	380-386	1999 Jun 3 22.39	1999 Jun 6 06.02
58	91.18	329-336	2000 May 26 23.03	2000 May 29 06.20
58	91.72	351-367		Permanent Component of ISS
58	91.67	348-364	2000 Nov 1 04.02	**2000 Nov 1 07.05
58	91.85	361-369	2000 Sep 18 03.44	2000 Sep 20 07.56
58	92.12	375-382	2000 Oct 20 15.08	2000 Oct 24 21.00
58	92.22	380-387	2001 Feb 24 10.06	
57	92.17	374-387	2001 Apr 18 12.40	
57	92.23	376-393	2001 May 6 02.21	2001 May 6 05.41
57	92.16	378-384	2000 Dec 1 16.23	
58	91.94	365-375	2001 Feb 8 11.26	**2001 Feb 8 13.50
58	92.09	374-380	2000 Dec 9 19.13	2000 Dec 11 23.03
57	91.67	350-363	*2001 Feb 16 14.06	2001 Feb 20 20.33
57	92.14	373-385	2001 Apr 16 08.48	**2001 Apr 16 13.23
				Permanent Component of ISS
57	92.07	371-381	2001 Mar 19 04.32	2001 Mar 21 07.31
57	92.22	375-392	2001 Apr 29 17.34	2001 May 1 16.11
57	92.40	380-404	2001 Oct 19 10.48	
64	92.47	390-401	2001 Oct 31 01.39	2001 Oct 31 04.59
57	92.29	375-399	2001 Aug 22 06.01	**2001 Aug 21
57	92.04	363-386	*2001 Jul 22 04.55	2001 Jul 25 03.39
				Permanent Component of ISS
63	92.40	385-399	2001 Aug 20 14.52	2001 Aug 22 18.23
64	92.48	390-402	2001 Nov 22 16.12	**2001 Nov 22
64	92.35	385-395	***2001 Sep 26 15.36	**2001 Sep 26 23.30
				Permanent Component of ISS
64	92.43	388-389		In Orbit at end of 2001
				In Orbit at end of 2001
64	92.09	370-384	2001 Dec 15 17.28	2001 Dec 17 17.55

e might differ slightly from the data shown in Table 1 for post-docking orbits and this is
mpared with the dates of the events to which they refer.

dule are indicated using the XYZ axes, with "L" indicating the port is along the

It had always been USSPACECOM's policy to catalogue everything in orbit which was large enough to be "seen" by its sensors and for which an orbit could be derived (even if that orbit would not be released for certain classified US payloads): in a few cases of short-lived objects, catalogue numbers and international designators would be assigned but no orbital data would be issued.

The first launch to be considered in terms of a cataloguing "anomaly" is that of the Kvant module [1] to expand the Mir base module into a modular space station. At launch, when the configuration of the assembly was not clear to western observers, two objects were initially catalogued: 1987-030A was Kvant itself and 1987-030B was the third (orbital) stage of the Proton-K launch vehicle. After the successful attachment of Kvant to the Mir base module, a service module separated from Kvant and was catalogued as 1987-030C: the Kvant module retained the designator 1987-030A. This policy seemed to be a reasonable one, although Kvant itself had never been in free flight: throughout its orbital lifetime it had been attached either to its service module or to the Mir base module (or both).

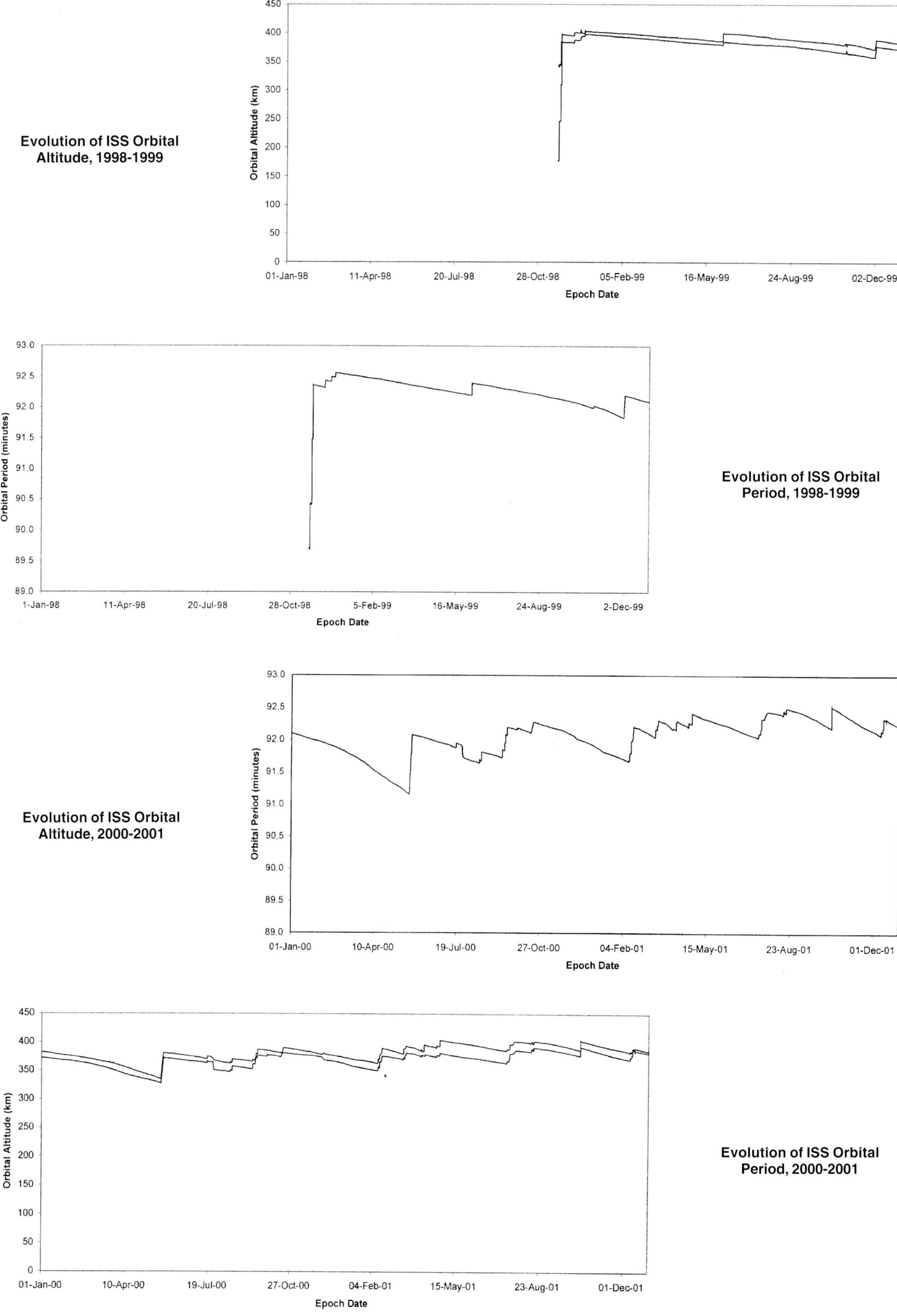

Evolution of ISS Orbital Altitude, 1998-1999

Evolution of ISS Orbital Period, 1998-1999

Evolution of ISS Orbital Altitude, 2000-2001

Evolution of ISS Orbital Period, 2000-2001

Although the scale was different, another mission to Mir saw this cataloguing policy reversed. In 1995 a visit of the US shuttle orbiter Atlantis to the Mir Complex carried a new docking module which would be used by future visiting shuttle orbiters. This module was carried in the orbiter's payload bay and then attached to the station's Kristall module on a permanent basis. Once more, the new module was never in free flight, but this time USSPACECOM did *not* catalogue the docking module – only the shuttle orbiter (1995-061A). The situation was – in principle – the same as the launch and docking of Kvant, with the orbiter Atlantis playing the part of the Kvant service module and the docking module being the equivalent of Kvant itself.

The evolution of ISS involves the launch of both modules which will be in "solo" free flight (Russian modules, launched atop the three-stage Proton-K) and modules which are never in "solo" free flight themselves but which are carried to ISS attached to another spacecraft and then attached to ISS, after which the "service module" (or equivalent) undocks and returns to Earth.

The launch of the first US module to ISS, Unity, followed the policy set by the Kvant launch. Unity was carried inside the payload bay of the shuttle orbiter Endeavour (1998-069A), and while still attached to the orbiter it docked with the ISS base module Zarya. USSPACECOM catalogued Unity as 1998-069F, although the module had never been in free flight.

The third module of ISS to be launched was Zvezda in 2000, and on this flight the situation was clear-cut. Zvezda flew as an independent satellite between launch by a three-stage Proton-K and its docking with ISS: the module was catalogued as 2000-037A.

TABLE 3: *Cataloguing Anomalies of Space Station Modules.*

Module	Launch Vehicle/Transport Craft	International Designator	Comments
Module Launch to Salyut 6			
Cosmos 1267	Proton-K	1981-039A	In free flight
Module Launches to Salyut 7			
Cosmos 1443	Proton-K	1983-013A	In free flight
Cosmos 1686	Proton-K	1985-086A	In free flight
Module Launches to the Mir Complex			
Kvant	Proton-K + Service Module	1987-030A	Never in free flight
Kvant 2	Proton-K	1989-093A	In free flight
Kristall	Proton-K	1990-048A	In free flight
Spektr	Proton-K	1995-024A	In free flight
Docking Module	Atlantis/STS-74	None	Never in free flight
Priroda	Proton-K	1996-023A	In free flight
Module Launches to the International Space Station			
Unity	Endeavour/STS-88	1998-069F	Never in free flight
Zvezda	Proton-K	2000-037A	In free flight
Destiny	Atlantis/STS-98	2001-006A	Never in free flight
Quest	Atlantis/STS-104	None	Never in free flight
Pirs	Soyuz-U + Service Module	None	Never in free flight

Table 3 shows a complete listing of all space station module launches, starting with Cosmos 1267 flown to Salyut 6 in 1981 and finishing with Pirs launched to ISS in 2001. It will be seen that when a module was in free flight USSPACECOM has *always* catalogued the module in its own right. The contradictory policy comes into play only when a module has *never* been in free flight. Of the modules never in free flight Kvant (Mir), Unity and Destiny (both ISS) have received their own catalogue designators, while the Docking Module (Mir), Quest and Pirs (both ISS) have not received their own designators.

Pirs, in particular, was essentially a repeat of the philosophy seen when Kvant was launched to Mir. The Pirs assembly was catalogued as 2001-041A after launch. Pirs itself was attached to a service module and this module was used to manoeuvre Pirs through to its docking with ISS. Two days after the Pirs assembly docked with ISS the service module was separated and de-orbited itself. In this instance (compare with the Kvant cataloguing policy) the service module retained the 2001-041A designator and Pirs has never received its own designator.

It is not the intention of this discussion to decide which of the two cataloguing policies is correct, but clearly USSPACECOM has shown itself to be inconsistent in the policy used for space station modules. A solution would be for USSPACECOM to "back-track" and retrospectively assign international designators and catalogue numbers to the modules not yet catalogued: one obvious problem with this solution is that the Docking Module attached to the Mir Complex is no longer in orbit – it came down with the Mir Complex on March 23, 2001.

Naming of ISS Missions

Presumably working on the basis that one can never have too many designators for space missions, NASA devised a new series of mission designators, covering both US and Russian launches within the ISS programme. These were presumably thought to be useful because they would relate to the mission being undertaken, rather than the almost-random international designator for the launch and the almost-as-random "STS" designators for missions which could get changed as shuttle flights were re-numbered. The designators used for the launches through to the end of 2001 are summarised in Table 4.

TABLE 4: *NASA Designators for ISS Missions.*

Spacecraft	International Designator	NASA Designator
ISS/Zarya	1998-067A	1R
Endeavour/STS-88	1998-069A	1A
Discovery/STS-96	1999-030A	2A.1
Atlantis/STS-101	2000-027A	2A.2
ISS/Zvezda	2000-037A	2R
Progress-M1 3	2000-044A	1P
Atlantis/STS-106	2000-053A	2A.2B
Discovery/STS-92	2000-062A	3A
Soyuz-TM 31	2000-070A	1S
Progress-M1 4	2000-073A	2P
Endeavour/STS-97	2000-078A	4A
Atlantis/STS-98	2001-006A	5A
Progress-M 44	2001-008A	3P
Discovery/STS-102	2001-010A	5A.1
Endeavour/STS-100	2001-016A	6A
Soyuz-TM 32	2001-017A	2S
Progress-M1 6	2001-021A	4P
Atlantis/STS-104	2001-028A	7A
Discovery/STS-105	2001-035A	7A.1
Progress-M 45	2001-036A	5P
Pirs	[2001-041]	SO1
Soyuz-TM 33	2001-048A	3S
Progress-M1 7	2001-051A	6P
Endeavour/STS-108	2001-054A	UF1

A problem has arisen in that some uninformed sections of the aerospace media have decided to use these purely-internal designators as the actual names of Russian spacecraft. For example, the (correctly-named) Soyuz-TM 32 mission has become "Soyuz 2" and Progress-M1 7 has become "Progress 6" in uninformed parlance. Of course, the designators in quotation marks are completely incorrect, but sadly NASA's public affairs department has done little to correct these regular and oft-quoted mis-namings of satellites.

One should bear in mind that NASA has no responsibility in naming other countries' satellites. If the Russians call their mission "Soyuz-TM 32", then that is the *correct* and *only* name which should be used when referring to the spacecraft. One wonders how NASA would react if Discovery/STS-105 were to become commonly called "STS-7", for example. "Soyuz 2" – as every space historian knows – was an unmanned spacecraft launched in October 1968 on an intended docking mission with the piloted Soyuz 3: and "Progress 6" was not launched to ISS in 2001 but was a cargo freighter launched to Salyut 6 in 1979.

There are some other examples of NASA-sourced misleading information being issued to the media. Perhaps the worst example was on the front page of the *official* NASA press kit for the Discovery/STS-96 mission: the thoughtless sub-title of the press kit, the second US launch in the ISS programme, was "First Visit to an Outpost in Orbit". Obviously, one asks "what about visits to Salyut, Skylab, Salyut 3, Salyut 4, Salyut 5, Salyut 6, Salyut 7 and the Mir Complex ?". Possibly without intention, the public relations bureaucrat who decided upon this title has erased the previous 28 years of Soviet/CIS space station operations, as well as the US missions to Skylab and the Mir Complex.

On a slightly-related matter (while dealing with mission classifications and types), it should be noted that there are generally three classes of crews trained for ISS missions.

One group comprises primarily NASA astronauts trained for shuttle launches and can include astronauts/ cosmonauts from other ISS partner countries and organisations (*eg*, Canada, France, ESA) who fly short-duration (~2 weeks) visits to ISS to deploy and/or integrate new components outside the station and also deliver supplies to the ISS resident crew.

A second group comprises both NASA astronauts and Russian cosmonauts training for the "resident" missions to ISS, these flights typically lasting for ~4-5 months (depending on changes to the shuttle launch schedules). The first ISS resident crew (ISS-1) was launched on Soyuz-TM 32 and they returned to Earth aboard a shuttle flight. Other ISS resident crews through to the end of 2001 have been both launched and recovered on shuttle flights (see Table 5).

TABLE 5: *Launches and Recoveries of Resident ISS Crews.*

Resident Crew Designator	Spacecraft at Launch	Spacecraft at Recovery
ISS-1	Soyuz-TM 31	Discovery/STS-102
ISS-2	Discovery/STS-102	Discovery/STS-105
ISS-3	Discovery/STS-105	Endeavour/STS-108
ISS-4	Endeavour/STS-108	[Endeavour/STS-111]

Note: The Endeavour/STS-111 mission was flown in June 2002.

The third group comprises cosmonauts and others who train for "Taxi Soyuz" missions. The Soyuz-TM spacecraft has a guaranteed "orbital lifetime" of about six months (individual Soyuz-T and –TM flights in the Soviet/ Russian domestic space programme have flown for longer than this), and therefore the Soyuz-TM emergency-return lifeboat needs to be replaced every six months or so. For such mission there is always a Russian Air Force mission commander, but the other one or two crewmembers can be from other ISS partner organisations (*eg*, ESA) or be purely commercial "space tourists". The crews for the Taxi Soyuz fly for about 8-12 days, being launched in a new spacecraft and then returning in the older spacecraft which had been already docked with ISS.

Debris From ISS

Any orbital station generates a debris cloud, no matter how careful crewmembers might be in disposing of rubbish. Even when the standard rubbish disposal *via* Progress freighters became the norm with the Mir Complex, there was a constant cloud of debris particles surrounding the station. These particles were too small to be catalogued individually, but represented paint flecks, particles of outside insulation material, *etc*. A similar cloud can be expected to evolve around ISS.

The larger pieces of "debris" which reach the official USSPACECOM catalogues (and some which remain in the classified 80,000 series catalogues for bureaucratic reasons) represent either ejected satellites, operational debris (the standard example would be covers placed over instrumentation at launch and then separated once in orbit), accidental "space junk" and then smaller pieces of the station or associated equipment. In the cases of flights to ISS there have been examples of both US and Russian spacecraft deploying small satellites once their work with ISS has been completed and the craft are on their return journey to Earth. Such small satellites are:

Endeavour/STS-88	1998-069B - SAC A
	1998-069C -MIGHTYSAT 1
Discovery/STS-96	1999-030B – Starshine
Discovery/STS-105	2001-035B – SIMPLESAT 1
Progress-M1 7	2001-051C – Kolibri
Endeavour/STS-108	2001-054B – Starshine 2

For Russian launches there are always initially two objects in orbit: the payload and the final stage of the launch vehicle which is either the Blok I third stage from the Soyuz-U and Soyuz-FG launch vehicles (the latter has been used on some Progress-M1 missions) or the 8S812 third stage from the Proton-K launch vehicle. Naturally, the rocket stages can be classified as being "operational debris".

By their very nature, upon orbital injection the space shuttle missions only have the one object in orbit – the orbiter itself: there is no rocket stage to separate.

Table 6 provides a listing of the debris which have been catalogued in addition to the small satellites and rocket stages which have just been noted. While the objects are apparently randomly assigned between ISS itself and the shuttle mission docked with the station at the time that the debris appeared, all of the objects can be generally classified as being "ISS debris". For convenience, the objects are listed in Table 6 in the order that they were catalogued (*ie*, by catalogue *number* which is not listed in the Table) rather than by the international designator of the mission to which they are assigned.

TABLE 6: *Debris from ISS and Associated Missions.*

Parent Mission	International Designator	Orbit Epoch Perigee			Incl °	Period min	Altitude km	Arg of °	Decay Date			Comments
Endeavour/STS-88	1998-069D	1998	Dec	8.25	51.60	92.32	382-394	280	1999	Apr	8	Slidewire carrier
Endeavour/STS-88	1998-069E	1998	Dec	8.64	51.61	92.31	382-394	315	1999	Mar	24	Socket (+ Tether?)
ISS/Zarya	1998-067C	1998	Dec	10.69	51.60	92.27	380-392	240	1998	Dec	14	Insulation blanket
ISS/Zarya	1998-067D	1999	Jul	4.79	51.59	91.60	351-355	222	1999	Aug	18	
Atlantis/STS-106	2000-053B	2000	Sep	13.42	51.58	91.46	339-354	24	2000	Sep	20	EVA on Sep 11
Atlantis/STS-106	2000-053C	2000	Sep	14.38	51.58	91.47	341-353	34	2000	Sep	25	EVA on Sep 11
ISS/Zarya	1998-067E	2000	Oct	18.70	51.58	91.89	365-370	327	2000	Nov	1	EVAs on Oct 17, 18
Discovery/STS-102	2001-010B	2001	Mar	15.02	51.57	92.03	371-376	210	2001	Jun	17	PAD
ISS/Zarya	1998-067F	2001	Oct	10.26	51.63	89.88	268-277	182	2001	Oct	10	EVA on Oct 8
ISS/Zarya	1998-067G	2001	Oct	16.75	51.63	90.67	307-314	234	2001	Oct	18	EVAs on Oct 8, 15
ISS/Zarya	1998-067H	2001	Oct	16.88	51.64	90.91	319-327	211	2001	Oct	19	EVAs on Oct 8, 15
ISS/Zarya	1998-067J	2001	Oct	16.02	51.62	90.10	279-291	195	2001	Oct	17	EVAs on Oct 8, 15
ISS/Zarya	1998-067K	2001	Oct	11.67	51.66	90.48	298-315	188	2001	Oct	12	EVA on Oct 8
ISS/Zarya	1998-067L	2001	Oct	16.20	51.64	92.45	394-400	207	2001	Oct	26	EVAs on Oct 8, 15
ISS/Zarya	1998-067M	2001	Oct	16.96	51.64	92.36	390-396	212	2001	Oct	26	EVAs on Oct 8, 15

It is possible to equate a few pieces of debris with objects which are known to have been metaphorically "dropped" in orbit (remembering that this is a zero-g environment!).

While an EVA was being undertaken on December 14, 1998 from the Endeavour/STS-88 mission, two small objects were "lost overboard". One was a slidewire carrier, 0.14 x 0.24 x 0.42 metres with a mass of 3.9 kg, and this was catalogued as 1998-061D (*ie*, assigned to the shuttle orbiter). Another object was a workstation interface socket, 0.09 x 0.10 x 0.13 metres, mass 1.1 kg: this was catalogued as 1998-061E. A tether was also "lost" during the EVA: ~0.16 x 0.05 x 0.03 metres, with a mass of <0.5 kg, it might have been attached to the socket. During the mission's second EVA an insulation blanket drifted away from ISS/Zarya, and this was catalogued as 1998-067C (*ie*, assigned to ISS/Zarya).

While the embryonic ISS, comprising only the Zarya and Unity modules, was in solo flight, a new piece of debris was catalogued with the first set of orbital data being issued for July 4, 1999. This was a month after the Discovery/STS-96 visit to ISS, and therefore it is difficult to say what this piece of debris represented. Maybe it had separated during the joint activities with Discovery and there had been a delay in cataloguing it ? The decay rate suggests that this might not be the case. Maybe it was simply something like a small piece of insulation material which drifted away from ISS while it was unoccupied?

The remaining pieces of ISS-related debris appear to be linked with EVA work by visiting shuttle crews. Two new pieces of debris assigned to the Atlantis/STS-106 mission were catalogued shortly after EVA operations on September 11, 2000. Another piece – this time assigned as ISS debris – appeared on October 18, 2000, following EVAs on October 17 and 18.

During an EVA in March 2001 a "PAD" which was used to attach equipment to the remote manipulator went missing and was catalogued as Discovery/STS-102 debris, 2001-010B.

The final debris events through to the end of 2001 appear to be associated with EVAs conducted on October 8 and 15, 2001. Seven new pieces of debris were catalogued as ISS debris during October 10-16, and they were in rapidly-decaying orbits. It is most likely that they originated during the October 8 EVA, but some could have been from the EVA on October 15.

Acknowledgement

The writer would like to thank the staff at the GSFC Orbit Information Group for the continued supply of Two-Line Orbital Elements, without which this review would have been impossible.

References

1. The first Salyut space station launched in April 1971 was simply named "Salyut" by the Russians, but as later Salyut stations appeared – Salyut 2, Salyut 3, *etc* - western observers added the sequential number "1" to the first Salyut's name. The Russians have *never* used the "Salyut 1" designator – the first station has always been "Salyut". Similarly, the first module added to the Mir base block has always been called simply "Kvant" by the Russians, but when Kvant 2 was launched to expand the orbital complex western writers similarly decided that the first Kvant should be called "Kvant 1". Once more, the Russians have never used the "Kvant 1" name.

2. The URL ("address") of the NASA GSFC/OIG web site is oig1.gsfc.nasa.gov.

*   *   *

The International Space Station
From Imagination to Reality

CONCLUSION

We hope you enjoyed reading the first of what we hope will be a series of volumes on the contrsuction of ISS. The British Interplanetary Society intends to publish a volume on the ISS every 3 or 4 years. It will depend on the speed of construction.

I wish to thank the contributors: Phillip Clark, Bart Hendrickx, Neville Kidger, Andrew Salmon, Roelof Schuiling, David J. Shayler and Bert Vis for all their hard work. Suszann Parry and Ben Jones from the BIS for their help and assistance. Jody Russell from the NASA stills oface. The staff of Novosti cosmonautika.

If you have any comments on this publication I would appreciate hearing from you. Thank you for your interest in Spaceflight and supporting the work of the The British Interplanetary Society.

Rex Hall

A view through Endeavour's aft flight deck of the International Space Station following undocking of STS-108.   (NASA)